TEST PREP

SPORT PILOT

STUDY AND PREPARE FOR THE SPORT PILOT AIRPLANE, LIGHTER-THAN-AIR, GLIDER, POWERED PARACHUTE, WEIGHT-SHIFT CONTROL, AND GYROPLANE FAA KNOWLEDGE EXAMS

- STUDY CURRENT INFORMATION WITH FREE ONLINE UPDATES AND FREE EMAIL SUBSCRIPTION SERVICE
- ANSWERS, EXPLANATIONS, REFERENCES AND ADDITIONAL STUDY MATERIAL INCLUDED FOR EACH CHAPTER
- INCLUDES THE OFFICIAL FAA COMPUTER TESTING SUPPLEMENT
- PLUS…HELPFUL TIPS AND INSTRUCTIONS FOR THE FAA KNOWLEDGE EXAM
- ORGANIZED BY SUBJECT FOR LOGICAL STUDY

READER TIP: THE FAA KNOWLEDGE EXAM QUESTIONS CAN CHANGE UP TO 3 TIMES A YEAR. STAY CURRENT WITH ALL TEST CHANGES; SIGN UP FOR ASA'S FREE EMAIL UPDATE SERVICE AT **WWW.ASA2FLY.COM/TESTUPDATE**

Sport Pilot Test Prep
2009 Edition

Aviation Supplies & Academics, Inc.
7005 132nd Place SE
Newcastle, Washington 98059-3153
425.235.1500
www.asa2fly.com

Important: This Test Prep should be sold with and used in conjunction with *Computer Testing Supplement for Sport Pilot/Instructor* (FAA-CT-8080-10A).

ASA reprints the FAA test figures and legends contained within this government document. It is also sold separately and available from aviation retailers nationwide. Order #ASA-CT-8080-10A.

ASA-TP-SPORT-09
ISBN 1-56027-692-4
978-1-56027-692-0

Printed in the United States of America

09 08 5 4 3 2 1

About the Contributors

Paul Hamilton
Sport Pilot Expert
Adventure Productions

Paul has been flying for over 30 years and holds an Advanced Flight Instructor and Ultralight Flight Instructor–Examiner ratings with all the national ultralight organizations. He is also a Master-Rated Hang Glider Pilot, Advanced-Rated Paraglider Pilot, and holds a patent for a variable camber inflatable wing. Paul holds Sport Pilot and Instructor ratings, and is an FAA Designated Pilot Examiner. He also developed and taught hang gliding with the University of Nevada at Reno. He holds a B.S. Degree in Mechanical Engineering, specializing in Aeronautics. Paul worked as an engineer in Design, Analysis, and Flight Testing, developing the business turboprop jet. Paul is the founder of Adventure Productions, a full service multi-media production company specializing in adventure sport aviation productions. He is producer of over forty films on ultralights and light sport aircraft—fixed wing, trikes, powered paragliders, powered parachutes, paragliders, and hang gliders. Today, Paul specializes in weight-shift aircraft and operates a flight school bringing the freedom of flight to many.

Jackie Spanitz
Director of Curriculum Development
Aviation Supplies & Academics, Inc.

Jackie Spanitz earned a bachelor of science degree with Western Michigan University (WMU), in Aviation Technology and Operations—Pilot option. In her masters program at Embry-Riddle Aeronautical University, she earned a degree in Aeronautical Science, specializing in Management. As Director of Curriculum Development for ASA, Jackie oversees new and existing product development, ranging from textbooks and flight computers to flight simulation software products, and integration of these products into new and existing curricula.

About ASA: Aviation Supplies & Academics, Inc. (ASA) is an industry leader in the development and sale of aviation supplies and publications for pilots, flight instructors, flight engineers, air traffic controllers, flight attendants, and aviation maintenance technicians. We manufacture and publish more than 300 products for the aviation industry. Aviators are invited to call 1-800-ASA-2-FLY for a free copy of our catalog. Visit ASA on the web: **www.asa2fly.com**

Contents

Preface

Welcome to ASA's Test Prep Series. ASA's test books have been helping pilots prepare for the FAA Knowledge Tests since 1984 with great success. We are confident that with proper use of this book, you will score very well on the sport pilot certificate test.

The FAA Sport Pilot questions have been arranged into chapters based on subject matter. Topical study, in which similar material is covered under a common subject heading, promotes better understanding, aids recall, and thus provides a more efficient study guide. Sport Pilot Instructors and Sport Pilot Examiners should use the *Certified Flight Instructor Test Prep* (ASA-TP-CFI) to prepare for your tests.

It is important to answer every question assigned on your FAA Knowledge Test. If in their ongoing review, the FAA authors decide a question has no correct answer, is no longer applicable, or is otherwise defective, your answer will be marked correct no matter which one you chose. However, you will not be given the automatic credit unless you have marked an answer. Unlike some other exams you may have taken, there is no penalty for "guessing" in this instance.

The FAA does not supply the correct answers to questions reproduced in this book, is not responsible for answers contained herein, and will not reveal what they consider the correct answers to be. The question and answer choices come from the FAA question bank; however, the FAA presents the questions in a different numerical sequence, they change the sequence of the A, B, C answer choices on the FAA website (http://www.faa.gov/pilots/testing/), and they include only samples of typical questions. They do this to discourage applicants from learning the test material by rote memory. The ASA test preps include a much wider sampling of the questions you may be asked at the test centers. A clear explanation is given directly below each question. Be careful to fully understand the intent of each question and corresponding answer while studying, rather than memorize the A, B, C answer. You may be asked a question that does not look familiar; studying and understanding the information in this book will give you the tools to answer these questions with confidence. If your study leads you to question an answer choice, we recommend you seek the assistance of a local ground or flight instructor. If you still believe the answer needs review, please forward your questions, recommendations, or concerns to:

Aviation Supplies & Academics, Inc.
7005 132nd Place SE
Newcastle, WA 98059-3153

Voice: 425.235.1500
Fax: 425.235.0128

Website: www.asa2fly.com
E-mail: CFI@asa2fly.com

Technical Editor:
Jackie Spanitz, ASA

Editor:
Jennie Trerise, ASA

Production Editor:
Cynthia Wyckoff, ASA

The FAA appreciates testing experience feedback. You can contact the branch responsible for the FAA Knowledge Exams directly at:

Federal Aviation Administration
AFS-630, Airman Testing Standards Branch
PO Box 25082
Oklahoma City, OK 73125
E-mail: AFS630comments@faa.gov

Updates and Practice Tests

Free Test Updates for the One-Year Lifecycle of Test Prep Books

The FAA releases a new test database each June, and makes amendments to this database approximately twice a year. However, a small number of questions may be withheld from the public for a period of time while the FAA gathers statistics and validates these questions. This means not all the questions are available to the public via the internet-posted databases, but they are being issued at the FAA testing centers. ASA combines years of experience with expertise in working with the tests to prepare the most comprehensive and accurate test preparation materials available in the industry.

You can feel confident that you will be prepared for your FAA Knowledge Exam by using ASA Test Prep products. ASA publishes test books each July and keeps abreast of all changes to the tests, as well as new questions that have been validated. These changes are then posted on the ASA website as a Test Update.

Visit the ASA website before taking your test to be certain you have all the current information. In addition, sign up for ASA's **free email Update service**. We will then send you an email notification if there is a change to the test you are preparing for so you can review the Update for revised and/or new test information.

www.asa2fly.com/testupdate

asa

Practice exams available for:

Sport Pilot • Private Pilot
Instrument Rating
Commercial Pilot
Flight Instructor
Ground Instructor
Fundamentals of Instructing
Flight Engineer
Airline Transport Pilot
General • Airframe • Powerplant

Prepware.com

Simulated testing center lets you practice for writtens!

As the experts in FAA Knowledge Exam preparation, we want you to have the confidence needed before heading to the testing center, and help eliminate the hassle and expense of retaking exams.

At **Prepware.com** you will experience:

- **Realistic Test Simulation**—Actual FAA test questions, in test form and time allowed for exam
- **Performance Graphs**—Review how you did, track your performance and review explanations for the questions you missed
- **Get Your Sign-Off**—Simply email your practice test results to ASA, and receive your instructor endorsement and graduation certificate (2 out of 5 practice tests with scores of at least 80%)
- **Gain Confidence**—Go into your exam fully prepared after practicing up to 5 simulated tests

Visit www.prepware.com for a free Demo!

Description of the Tests

All test questions are the objective, multiple-choice type, with three choices of answers. Each question can be answered by the selection of a single response. Each test question is independent of other questions, that is, a correct response to one does not depend upon or influence the correct response to another.

For the most efficient and effective study program, refer to the table below and place emphasis on those questions most likely to be included on your test (identified by the aircraft category above each question number).

Test Code	Test Name	Test Prep Study	Number of Questions	Age	Allotted Time
SPA	Sport Pilot Airplane	ALL, AIR	40	15	2.0
SPB	Sport Pilot Lighter-Than-Air (Balloon)	ALL, LTA	40	15	2.0
SPI	Sport Pilot Glider	ALL, GLI	40	15	2.0
SPL	Sport Pilot Lighter-Than-Air (Airship)	ALL, LTA	40	15	2.0
SPP	Sport Pilot Powered Parachute	ALL, PPC	40	15	2.0
SPW	Sport Pilot Weight-Shift Control	ALL, WSC	40	15	2.0
SPY	Sport Pilot Gyroplane	ALL, RTC	40	15	2.0

A score of 70 percent must be attained to successfully pass each test.

Sport Pilot Instructors should use the *CFI Test Prep* (#ASA-TP-CFI) to prepare for their test.

Acceptable Forms of Authorization

- Certificate of graduation or a statement of accomplishment certifying the satisfactory completion of the ground school portion of a course for the certificate or rating sought. The certificate or statement may be issued by a Federal Aviation Administration certified pilot school or an agency such as a high school, college, adult education program, Civil Air Patrol or Reserve Officers Training Corp (ROTC) flight training school.
- Written statement or logbook endorsement from an authorized ground or flight instructor certifying that the applicant completed an applicable ground training or home study course and is prepared for the knowledge test.
- Failed, passing or expired Airman Knowledge Test Report, provided the applicant still has the ***original*** test report in his/her possession. (*See* the following **Retesting** explanations.)
- "Expired test/credit" letter issued by the Airmen Certification Branch (in lieu of a duplicate Airman Knowledge Test Report).
- The Prepware software (ASA-TW-SPORT) allows you to study, practice taking tests, and then email the results to receive your authorization from ASA.
- The online simulated testing center **www.prepware.com** lets you practice taking tests and receive your endorsement from ASA.

When you go to take your FAA Knowledge Test, you will be required to show proper identification and have certification of your preparation for the examination, signed by an authorized Flight or Ground Instructor. Ground Schools will have issued the endorsements as you complete the course. If you choose a home-study for your Knowledge Test, you can get either an endorsement from your instructor or submit your home-study materials to an FAA Office for review and approval prior to taking the test.

Sport Pilot Knowledge Test Endorsement

I certify that (First name, MI, Last name) ______________________________

has received the required aeronautical knowledge training of 14 CFR §61.309. I have determined he/she is prepared for the Sport Pilot (category) ____________________ knowledge test.

Signed ______________________________ Date ____________________

CFI Number ______________________________ Expires ____________________

Retesting Procedures

- Applicants retesting ***after failure*** are required to submit the applicable score report indicating failure, along with an endorsement from an authorized instructor who gave the applicant the additional training, and certifying the applicant is competent to pass the test. The original failed test report presented as authorization must be retained by the proctor and attached to the applicable sign-in/out log. The latest test taken will reflect the official score.
- Applicants retesting ***in an attempt to achieve a higher passing score*** may retake the same test for a better grade after 30 days. The latest test taken will reflect the official score. Applicants are required to submit the ***original*** applicable score report indicating previous passing score to the testing center prior to testing. Testing center personnel must collect and destroy this report prior to issuing the new test report.
- Applicants approaching the 24-month deadline can use their previous test results as authorization as long as the test results have not expired. After the 24-month period, you must receive authorization as if you were starting from scratch.

Process for Taking a Knowledge Test

The Federal Aviation Administration (FAA) has available hundreds of computer testing centers worldwide. These testing centers offer the full range of airman knowledge tests including military competence, instrument foreign pilot, and pilot examiner predesignated tests. Refer to the list of computer testing designees (CTDs) at the end of this section.

The first step in taking a knowledge test is the registration process. You may either call the testing centers' 1-800 numbers or simply take the test on a walk-in basis. If you choose to use the 1-800 number to register, you will need to select a testing center, schedule a test date, and make financial arrangements for test payment. You may register for tests several weeks in advance, and you may cancel your appointment according to the CTD's cancellation policy. If you do not follow the CTD's cancellation policies, you could be subject to a cancellation fee.

The next step in taking a knowledge test is providing proper identification, typically a valid driver's license and test authorization. Testing center personnel will not begin the test until proper identification and authorization are verified.

Upon completion of the knowledge test, you will receive your Airman Test Report, with the testing center's embossed seal, which reflects your score.

The Airman Test Report lists the learning statement codes for questions answered incorrectly. The total number of learning statement codes shown on the Airman Test Report is not necessarily an indication of the total number of questions answered incorrectly. Study these knowledge areas to improve your understanding of the subject matter. *See* the *Learning Statement Code/Question Number Cross-Reference* in the back of this book for a complete list of which questions apply to each learning statement code.

Your instructor is required to provide instruction on each of the knowledge areas listed on your Airman Test Report and to complete an endorsement of this instruction prior to taking the practical test. During the oral portion of the practical test, the examiner is required to evaluate the noted areas of deficiency.

Should you require a duplicate Airman Test Report due to loss or destruction of the original, send a signed request accompanied by a check or money order for $1 payable to the FAA. Your request should be sent to the Federal Aviation Administration, Airmen Certification Branch, AFS-760, P.O. Box 25082, Oklahoma City, OK 73125.

Airman Knowledge Testing Sites

The following is a list of the computer testing designees authorized to give FAA knowledge tests. This list should be helpful in case you choose to register for a test or simply want more information. The latest listing of computer testing center locations is available on the FAA website at http://www.faa.gov/pilots/testing, under "Knowledge Test Centers" select "Center List" and a PDF will download automatically.

Computer Assisted Testing Service (CATS)
1801 Murchison Drive, Suite 288
Burlingame, CA 94010
Applicant inquiry and test registration: 1-800-947-4228
From outside the U.S.: (650) 259-8550

LaserGrade Computer Testing
16821 S.E. McGillivray, Suite 201
Vancouver, WA 98683
Applicant inquiry and test registration: 1-800-211-2754
From outside the U.S.: (360) 896-9111

International pilots who want to apply for an FAA certificate based on their ICAO foreign certificates should go to the nearest FAA Flight Standards District Office (FSDO). Phone numbers for these offices will be found in the blue pages of the local telephone book. This will allow you to fly a U.S.-registered aircraft while in the U.S. If you hold instrument privileges on your foreign license, you can take a 50-question knowledge test (Instrument Rating—Foreign Pilot), and if you pass it, have instrument privileges added to this FAA private pilot certificate.

If you are outside of the U.S., you will have to go in person to an FAA International Field Office (IFO) and apply for an FAA private pilot certificate. When outside of the U.S. you will only be authorized to fly a U.S.-registered aircraft.

International Field Offices (IFO)

1. Brussels, Belgium (32-2) 508.2721
   FAA C/O American Embassy, PSC 82 Box 002, APO AE 09710
2. Frankfurt, Germany (49-69) 69.705.111
   FAA C/O IFO EA-33 Unit 7580, APO AE 09050
3. London, England (44-181) 754.88.19
   FAA C/O American Embassy, PSC 801 Box 63, FPO AE 09498-4063
4. Singapore (65) 543-1466
   FAA C/O American Embassy, PSC 470 AP 96507-0001

These special certificates will not allow you to fly for hire in the U.S. To qualify for a "clean" FAA commercial pilot certificate or higher, you must meet the full certification requirements of 14 CFR Part 61, for the level of certificate you are requesting. Your current, logged flying time will count towards the required experience. However, all required training, knowledge, and practical tests must be completed.

Use of Test Aids and Materials

Airman knowledge tests require applicants to analyze the relationship between variables needed to solve aviation problems, in addition to testing for accuracy of a mathematical calculation. The intent is that all applicants are tested on concepts rather than rote calculation ability. It is permissible to use certain calculating devices when taking airman knowledge tests, provided they are used within the following guidelines. The term "calculating devices" is interchangeable with such items as calculators, computers, or any similar devices designed for aviation-related activities.

1. Guidelines for use of test aids and materials. The applicant may use test aids and materials within the guidelines listed below, if actual test questions or answers are not revealed.
   a. Applicants may use test aids, such as scales, straightedges, protractors, plotters, navigation computers, log sheets, and all models of aviation-oriented calculating devices that are directly related to the test. In addition, applicants may use any test materials provided with the test.
   b. Manufacturer's permanently inscribed instructions on the front and back of such aids listed in 1(a), e.g., formulas, conversions, regulations, signals, weather data, holding pattern diagrams, frequencies, weight and balance formulas, and air traffic control procedures are permissible.
   c. The test proctor may provide calculating devices to applicants and deny them use of their personal calculating devices if the applicant's device does not have a screen that indicates all memory has been erased. The test proctor must be able to determine the calculating device's erasure capability. The use of calculating devices incorporating permanent or continuous type memory circuits without erasure capability is prohibited.
   d. The use of magnetic cards, magnetic tapes, modules, computer chips, or any other device upon which prewritten programs or information related to the test can be stored and retrieved is prohibited. Printouts of data will be surrendered at the completion of the test if the calculating device used incorporates this design feature.
   e. The use of any booklet or manual containing instructions related to the use of the applicant's calculating device is not permitted.
   f. Dictionaries are not allowed in the testing area.
   g. The test proctor makes the final determination relating to test materials and personal possessions that the applicant may take into the testing area.

2. Guidelines for dyslexic applicant's use of test aids and materials. A dyslexic applicant may request approval from the local Flight Standards District Office (FSDO) to take an airman knowledge test using one of the three options listed in preferential order:
   a. Option One. Use current testing facilities and procedures whenever possible.
   b. Option Two. Applicants may use Franklin Speaking Wordmaster® to facilitate the testing process. The Wordmaster® is a self-contained electronic thesaurus that audibly pronounces typed in words and presents them on a display screen. It has a built-in headphone jack for private listening. The headphone feature will be used during testing to avoid disturbing others.
   c. Option Three. Applicants who do not choose to use the first or second option may request a test proctor to assist in reading specific words or terms from the test questions and supplement material. In the interest of preventing compromise of the testing process, the test proctor should be someone who is non-aviation oriented. The test proctor will provide reading assistance only, with no explanation of words or terms. The Airman Testing Standards Branch, AFS-630, will assist in the selection of a test site and test proctor.

Cheating or Other Unauthorized Conduct

Computer testing centers must follow strict security procedures to avoid test compromise. These procedures are established by the FAA and are covered in FAA Order 8080.6 Conduct of Airman Knowledge Tests. The FAA has directed testing centers to terminate a test at any time a test proctor suspects a cheating incident has occurred. An FAA investigation will then be conducted. If the investigation determines that cheating or unauthorized conduct has occurred, then any airman certificate or rating that you hold may be revoked, and you will be prohibited for 1 year from applying for or taking any test for a certificate or rating under 14 CFR Part 61.

Validity of Airman Test Reports

Airman Test Reports are valid for the 24-calendar month period preceding the month you complete the practical test. If the Airman Test Report expires before completion of the practical test, you must retake the knowledge test.

Eligibility Requirements

Always check the current 14 CFR Part 61 for pilot certificate requirements. To be eligible for a Sport Pilot Certificate a person must:

1. Be at least 17 years old (16 if you are applying to operate a glider or balloon).
2. Be able to read, speak, write, and understand English or have a limitation placed on the certificate.
3. Score at least 70 percent on the required FAA Knowledge Test.
4. Pass a practical test on the subjects and maneuvers outlined in the Sport Pilot Practical Test Standards (ASA-8081-SPORT *Sport Pilot for Airplane, Weight-Shift Control, Powered Parachute, and Flight Instructor*).
5. See 14 CFR §61.329 for the special provisions for obtaining a Sport Pilot certificate as a registered ultralight pilot with an FAA-recognized ultralight organization.
6. The table on the next page explains the aeronautical experience you must have to apply for a Sport Pilot Certificate.

Aeronautical Experience for the Sport Pilot Certificate

If you are applying for a sport pilot certificate with...	Then you must log at least...	Which must include at least...
(a) Airplane category and single-engine land or sea class privileges,	(1) 20 hours of flight time, including at least 15 hours of flight training from an authorized instructor in a single-engine airplane and at least 5 hours of solo flight training in the areas of operation listed in §61.311,	(i) 2 hours of cross-country flight training, (ii) 10 takeoffs and landings to a full stop (with each landing involving a flight in the traffic pattern) at an airport; (iii) One solo cross-country flight of at least 75 nautical miles total distance, with a full-stop landing at a minimum of two points and one segment of the flight consisting of a straight-line distance of at least 25 nautical miles between the takeoff and landing locations, and (iv) 3 hours of flight training on those areas of operation specified in §61.311 preparing for the practical test within 60 days before the date of the test.
(b) Glider category privileges, and you have not logged at least 20 hours of flight time in a heavier-than-air aircraft,	(1) 10 hours of flight time in a glider, including 10 flights in a glider receiving flight training from an authorized instructor and at least 2 hours of solo flight training in the areas of operation listed in §61.311,	(i) Five solo launches and landings, and (ii) 3 hours of flight training on those areas of operation specified in §61.311 preparing for the practical test within 60 days before the date of the test.
(c) Glider category privileges, and you have logged 20 hours flight time in a heavier-than-air aircraft,	(1) 3 hours of flight time in a glider, including five flights in a glider while receiving flight training from an authorized instructor and at least 1 hour of solo flight training in the areas of operation listed in §61.311,	(i) Three solo launches and landings, and (ii) 3 hours of flight training on those areas of operation specified in §61.311 preparing for the practical test within 60 days before the date of the test.
(d) Rotorcraft category and gyroplane class privileges,	(1) 20 hours of flight time, including 15 hours of flight training from an authorized instructor in a gyroplane and at least 5 hours of solo flight training in the areas of operation listed in §61.311,	(i) 2 hours of cross-country flight training, (ii) 10 takeoffs and landings to a full stop (with each landing involving a flight in the traffic pattern) at an airport, (iii) One solo cross-country flight of at least 50 nautical miles total distance, with a full-stop landing at a minimum of two points, and one segment of the flight consisting of a straight-line distance of at least 25 nautical miles between the takeoff and landing locations, and (iv) 3 hours of flight training on those areas of operation specified in §61.311 preparing for the practical test within 60 days before the date of the test.
(e) Lighter-than-air category and airship class privileges,	(1) 20 hours of flight time, including 15 hours of flight training from an authorized instructor in an airship and at least 3 hours performing the duties of pilot in command in an airship with an authorized instructor in the areas of operation listed in §61.311,	(i) 2 hours of cross-country flight training, (ii) Three takeoffs and landings to a full stop (with each landing involving a flight in the traffic pattern) at an airport, (iii) One cross-country flight of at least 25 nautical miles between the takeoff and landing locations, and (iv) 3 hours of flight training on those areas of operation specified in §61.311 preparing for the practical test within 60 days before the date of the test.
(f) Lighter-than-air category and balloon class privileges,	(1) 7 hours of flight time in a balloon, including three flights with an authorized instructor and one flight performing the duties of pilot in command in a balloon with an authorized instructor in the areas of operation listed in §61.311,	(i) 2 hours of cross-country flight training and (ii) 3 hours of flight training on those areas of operation specified in §61.311 preparing for the practical test within 60 days before the date of the test.
(g) Powered parachute category land or sea class privileges,	(1) 12 hours of flight time in a powered parachute, including 10 hours flight training and, and at least 2 hours solo flight training in the areas of operation listed in §61.311.	(i) 1 hour of cross-country flight training, (ii) 20 takeoffs and landings to a full stop in a powered parachute with each landing involving flight in the traffic pattern at an airport; (iii) 10 solo takeoffs and landings to a full stop (with each landing involving a flight in the traffic pattern) at an airport, (iv) One solo flight with a landing at a different airport and one segment of the flight consisting of a straight-line distance of at least 10 nautical miles between takeoff and landing locations, and (v) 3 hours of flight training on those areas of operation specified in §61.311 preparing for the practical test within 60 days before the date of the test.
(h) Weight-shift-control aircraft category land or sea class privileges,	(1) 20 hours of flight time, including 15 hours of flight training from an authorized instructor in a weight-shift-control aircraft and at least 5 hours of solo flight training in the areas of operation listed in §61.311,	(i) 2 hours of cross-country flight training; (ii) 10 takeoffs and landings to a full stop (with each landing involving a flight in the traffic pattern) at an airport, (iii) One solo cross-country flight of at least 50 nautical miles total distance, with a full-stop landing at a minimum of two points, and one segment of the flight consisting of a straight-line distance of at least 25 nautical miles between takeoff and landing locations, and (iv) 3 hours of flight training on those areas of operation specified in §61.311 preparing for the practical test within 60 days before the date of the test.

Test-Taking Tips

Follow these time-proven tips, which will help you develop a skillful, smooth approach to test-taking:

1. In order to maintain the integrity of each test, the FAA may rearrange the answer stems to appear in a different order on your test than you see in this book. For this reason, be careful to fully understand the intent of each question and corresponding answer while studying, rather than memorize the A, B, C answer choice.
2. You may be asked a question that does not look familiar; studying and understanding the information in this book will give you the tools to answer these questions with confidence.
3. Take with you to the testing center a sign-off from an instructor, photo I.D., the testing fee, calculator, flight computer (ASA's E6-B or CX-2 are both authorized for use on the test) and a plotter.
4. Your first action when you sit down should be to write on the scratch paper the weight and balance and any other formulas and information you can remember from your study. Remember, some of the formulas may be on your E6-B.
5. Answer each question in accordance with the latest regulations and guidance publications.
6. Read each question carefully before looking at the possible answers. You should clearly understand the problem before attempting to solve it.
7. After formulating an answer, determine which answer choice corresponds the closest with your answer. The answer chosen should completely resolve the problem.
8. From the answer choices given, it may appear that there is more than one possible answer. However, there is only one answer that is correct and complete. The other answers are either incomplete, erroneous, or represent popular misconceptions.
9. If a certain question is difficult for you, it is best to mark it for REVIEW and proceed to the other questions. After you answer the less difficult questions, return to those which you marked for review and answer them. Be sure to untag these questions once you have answered them. The review marking procedure will be explained to you prior to starting the test. Although the computer should alert you to unanswered questions, make sure every question has an answer recorded. This procedure will enable you to use the available time to the maximum advantage.
10. Perform each math calculation twice to confirm your answer. If adding or subtracting a column of numbers, reverse your direction the second time to reduce the possibility of error.
11. When solving a calculation problem, select the answer nearest to your solution. The problem has been checked with various types of calculators; therefore, if you have solved it correctly, your answer will be closer to the correct answer than any of the other choices.
12. Remember that information is provided in the FAA Figures and Legends in the Computer Testing Supplement, which will be given to you at the testing center to use during the test.
13. Remember to answer every question, even the ones with no completely correct answer, to ensure the FAA gives you credit for a bad question.
14. Take your time and be thorough but relaxed. Take a minute off every half-hour or so to relax the brain and the body. Get a drink of water halfway through the test.

Suggested Materials for the Sport Pilot Certificate

The following are some of the publications recommended for the Sport Pilot, Instructor, Examiner Certificates. All are reprinted by ASA and available from authorized ASA dealers and distributors. Be sure to use the latest revision of these references when reviewing for the test.

ASA-AC00-6A	*Aviation Weather*
ASA-AC00-45	*Aviation Weather Services*
ASA-8081-SPORT	*Sport Pilot PTS for Airplane, Weight-Shift Control, Powered Parachute, and Flight Instructor*
ASA-8083-1	*Aircraft Weight and Balance Handbook*
ASA-8083-3	*Airplane Flying Handbook*
ASA-8083-13	*Glider Flying Handbook*
ASA-8083-21	*Rotorcraft Flying Handbook*
ASA-8083-25	*Pilot's Handbook of Aeronautical Knowledge*
ASA-8083-29	*Powered Parachute Flying Handbook*
ASA-CP-RLX	Ultimate Rotating Plotter
ASA-CX-2	CX-2 Pathfinder Electronic Flight Computer
ASA-E6B	E6-B Flight Computer
ASA-FR-AM-BK	*Federal Aviation Regulations* and *Aeronautical Information Manual*
ASA-F2F-CKRIDE	*Sport Pilot Checkride*
ASA-F2F-CK-DVD	Sport Pilot Checkride DVD
ASA-F2F-LOG	Sport Pilot Logbook
ASA-F2F-FW-PF	*Preflight a Fixed Wing Light-Sport Aircraft for Sport Pilots*
ASA-F2F-FW-L2F	*Be a Sport Pilot: Learn to Fly a Fixed Wing Light-Sport Aircraft*
ASA-F2F-FW-KIT	Sport Pilot Fixed Wing Kit
ASA-F2F-SPORT	*Sport Pilot: Choosing the Light-Sport Aircraft That's Right for You*
ASA-F2F-W2F	*Weather to Fly for Sport Pilots*
ASA-TW-SPORT	*Prepware Exam Software for Sport Pilots*

Visit the ASA website for more information on all these products: **www.asa2fly.com**

ASA Test Prep Layout

The FAA questions have been sorted into chapters according to subject matter. Within each chapter, the questions have been further classified and all similar questions grouped together with a concise discussion of the material covered in each group. This discussion material of "Chapter text" is printed in a larger font and spans the entire width of the page. Immediately following the FAA Question is ASA's Explanation in *italics*. The last line of the Explanation contains the Learning Statement Code and further reference (if applicable). *See* the EXAMPLE below.

Figures referenced by the Chapter text only are numbered with the appropriate chapter number, i.e., "Figure 1-1" is Chapter 1's first chapter-text figure.

Some FAA Questions refer to Figures or Legends immediately following the question number, i.e., "3201. (Refer to Figure 14.)." These are FAA Figures and Legends which can be found in the separate booklet: *Computer Testing Supplement* (CT-8080-XX). This supplement is bundled with the Test Prep and is the exact material you will have access to when you take your computerized test. We provide it separately, so you will become accustomed to referring to the FAA Figures and Legends as you would during the test.

Figures referenced by the Explanation and pertinent to the understanding of that particular question are labeled by their corresponding Question number. For example: the caption "Questions 3245 and 3248" means the figure accompanies the Explanations for both Question 3245 and 3248.

Answers to each question are found at the bottom of each page.

EXAMPLE:

Chapter text

Four aerodynamic forces are considered to be basic because they act upon an aircraft during all flight maneuvers. There is the downward-acting force called WEIGHT which must be overcome by the upward-acting force called LIFT, and there is the rearward-acting force called DRAG, which must be overcome by the forward-acting force called THRUST.

Category rating. This question may be found on tests for these ratings.*

ALL, AIR, RTC, GLI, LTA, PPC, WSC

2201. (Refer to Figure 14.) The four forces acting on an airplane in flight are

See separate book: *Computerized Testing Supplement* (CT-8080-XX)

A— lift, weight, thrust, and drag.
B— lift, weight, gravity, and thrust.
C— lift, gravity, power, and friction.

FAA question and answer choices

Lift, weight, thrust, and drag are the four basic aerodynamic forces acting on an aircraft in flight.
(PLT247) — FAA-H-8083-25

Explanation

Answer (B) is incorrect because the force of gravity is always the same number and reacts with the airplane's mass to produce a different weight for almost every airplane. Answer (C) is incorrect because weight is the final product of gravity, thrust is the final product of power, and drag is the final product of friction. Power, gravity, and friction are only parts of the aerodynamic forces of flight.

Code line. FAA's Learning Statement Code in parentheses, followed by further references where applicable.

Incorrect answer explanation. Reasons why answer choices are incorrect explained here.

* **Note:** The FAA does *not* identify which questions are on the different ratings' tests. Unless the wording of a question is pertinent to only one rating category, it may be found on *any* of the tests.

ALL = All aircraft	AIR = Airplane	RTC = Rotorcraft (gyroplane)	PPC = Powered Parachute
GLI = Glider	WSC = Weight-Shift Control	LTA = Lighter-than-Air (balloon)	

Chapter 1
Regulations

Introduction

In addition to the popular reference "FAR" (which means Federal Aviation Regulations), the FAA cites these federal aviation regulations with reference to Title 14 of the Code of Federal Regulations. For example, "FAR Part 91.3" is referenced as "14 CFR Part 91 Section 3."

While **Federal Aviation Regulations** are many and varied, some are of particular interest to all pilots.

14 CFR Part 1 contains definitions and abbreviations of many terms commonly used in aviation. For example, the term "night" means "the time between the end of evening civil twilight and the beginning of morning civil twilight, as published in the American Air Almanac, converted to local time." This is the period of time during which a Sport Pilot may not fly as pilot-in-command of an aircraft.

14 CFR Part 21, entitled "Certification Procedures for Products and Parts" prescribes the procedural requirements for type certificates, airworthiness certificates, and parts.

14 CFR Part 43, entitled "Maintenance, Preventative Maintenance, Rebuilding, and Alteration" governs U.S. registered and operated aircraft and parts.

14 CFR Part 45, entitled "Identification and Registration Marking" governs U.S. registered and operated aircraft.

14 CFR Part 61, entitled "Certification: Pilots, Flight Instructors and Ground Instructors," prescribes the requirements for issuing pilot and flight instructor certificates and ratings, the conditions of issue, and the privileges and limitations of those certificates and ratings.

14 CFR Part 71 concerns airspace and lists specific airport designations ("B," "C," or "D").

14 CFR Part 91, entitled "General Operating and Flight Rules," describes rules governing the operation of aircraft (with certain exceptions) within the United States.

The **National Transportation Safety Board (NTSB)** has established rules and requirements for notification and reporting of aircraft accidents and incidents. These are contained in **NTSB Part 830**.

ALL

2124. The definition of nighttime is

A—sunset to sunrise.
B—1 hour after sunset to 1 hour before sunrise.
C—the time between the end of evening civil twilight and the beginning of morning civil twilight.

Night is the time between the end of evening civil twilight and the beginning of morning civil twilight converted to local time, as published in the American Air Almanac. (PLT395) — 14 CFR §1.1

Answer (A) is incorrect because it refers to the time when lighted position lights are required. Answer (B) is incorrect because it refers to the currency requirement for Private Pilots and above to carry passengers at night.

Answers
2124 [C]

Sport Pilot Certificate Privileges and Limitations

Sport Pilots **may** act as pilot in command of a light-sport aircraft. Sport Pilots may share the operating expenses of a flight with a passenger, provided the expenses involve only fuel, oil, airport expenses, or aircraft rental fees. You must pay at least half the operating expenses of the flight.

Sport Pilots **may not** act as pilot-in-command of a light-sport aircraft:

1. To carry a passenger or property for compensation or hire.
2. For compensation or hire.
3. In furtherance of a business.
4. While carrying more than one passenger.
5. At night.
6. In Class A airspace.
7. In Class B, C, and D airspace, at an airport located in Class B, C, or D airspace, and to, from, through, or at an airport having an operational control tower (unless you have further training and an endorsement to do so).
8. Outside the United States, unless you have prior authorization from the country in which you seek to operate. Your sport pilot certificate carries the limit "Holder does not meet ICAO requirements."
9. To demonstrate the aircraft in flight to a prospective buyer if you are an aircraft salesperson.
10. In a passenger-carrying airlift sponsored by a charitable organization.
11. At an altitude of more than 10,000 feet MSL.
12. When the flight or surface visibility is less than 3 statute miles.
13. Without visual reference to the surface.
14. If the aircraft has a V_H that exceeds 87 knots CAS (unless you have further training and an endorsement to do so).
15. Contrary to any operating limitation placed on the airworthiness certificate of the aircraft being flown.
16. Contrary to any limit or endorsement on your pilot certificate, driver's license (or airman medical certificate), or any other limit or endorsement from an authorized instructor.
17. Contrary to any restriction or limitation on your U.S. driver's license or any restriction or limitation imposed by judicial or administrative order when using your driver's license to satisfy a requirement of this part.
18. While towing any object.
19. As a pilot flight crewmember on any aircraft for which more than one pilot is required by the type certificate of the aircraft or the regulations under which the flight is conducted.

Pilot Ratings

Sport Pilot certificates do not list category or class — these are included in the logbook endorsement. *See* Figure 1-1A, Figure 1-1B, and Figure 1-1C.

When a private pilot certificate and above is issued, it lists the category, class, and type (if appropriate) of aircraft in which the certificate holder is qualified.

The term "**category**" means a broad classification of aircraft, such as airplane, rotorcraft, glider, weight-shift control, powered parachute, and lighter-than-air. The term "**class**" means a classification within a category having similar operating characteristics, such as single-engine, multi-engine, land and water. The term "**type**" means a specific make and basic model of aircraft, such as F-27, or DC-7.

A type rating must be held by the pilot-in-command of a large aircraft. "Large aircraft" means aircraft of more than 12 500 pounds maximum certificated takeoff weight.

All turbojet-powered airplanes, regardless of weight, require the PIC to have a type rating.

If a pilot changes his/her permanent mailing address without notifying the FAA Airman's Certification Branch, in writing, within 30 days, then he/she may not exercise the privileges of his/her certificate.

Figure 1-1A. Sport Pilot certificate (front)

Figure 1-1B. Sport Pilot certificate (back)

I certify that JACKIE SMITH
has met the training and endorsement requirements of 61.309, 61.311 and 61.313. I have determined him/her proficient to act as PIC of WSCL
COSMOS PHASE II Set WS-1
(make/ model) of light-sport aircraft.
5-14-2007
Paul D. Hamilton Date
DPE FS-60-0020 Exp date 9·30-2008 Pilot # 3051445

Figure 1-1C. Logbook endorsement

ALL
2130. If sunset is 2021 and the end of evening civil twilight is 2043, when must a sport pilot terminate the flight?

A—2021.
B—2043.
C—2121.

A sport pilot may not act as pilot-in-command of an aircraft at night. A sport pilot must land by the end of evening twilight. (PLT448) — 14 CFR §61.315

ALL
2123. How many passengers is a sport pilot allowed to carry on board?

A—One.
B—Two.
C—Three.

Sport Pilots may not act as pilot in command of a light-sport aircraft while carrying more than one passenger. (PLT448) — 14 CFR §61.315

ALL
2061. Outside controlled airspace, the minimum flight visibility requirement for a sport pilot flying above 1,200 feet AGL and below 10,000 feet MSL during daylight hours is

A—1 statute mile.
B—3 statute miles.
C—5 statute miles.

Minimum flight or surface visibility for sport pilots is 3 statute miles in all airspace at all times. (PLT467) — 14 CFR §61.315

ALL
2185. With respect to the certification of airmen, which is a category of aircraft?

A—Gyroplane, helicopter, airship, free balloon.
B—Airplane, rotorcraft, glider, lighter-than-air.
C—Single-engine land and sea, multiengine land and sea.

With respect to the certification of airmen, "category" means a broad classification of aircraft such as airplane, rotorcraft, glider, lighter-than-air, weight-shift control, and powered parachute. (PLT371) — 14 CFR §1.1

Answer (A) is incorrect because it refers to classes of rotorcraft and lighter-than-air craft. Answer (C) is incorrect because it refers to classes of airplanes.

ALL
2131. If a certificated pilot changes permanent mailing address and fails to notify the FAA Airmen Certification Branch of the new address, the pilot is entitled to exercise the privileges of the pilot certificate for a period of only

A—30 days after the date of the move.
B—60 days after the date of the move.
C—90 days after the date of the move.

The holder of a Pilot or Flight Instructor Certificate who has made a change in his/her permanent mailing address may not, after 30 days from the date moved, exercise the privileges of his/her certificate unless he/she has notified in writing the Department of Transportation, Federal Aviation Administration, Airmen Certification Branch, Box 25082, Oklahoma City, OK 73125, of the new address. (PLT387) — 14 CFR §61.60

Answers

2130 [B]	2123 [A]	2061 [B]	2185 [B]	2131 [A]

Responsibility and Authority of the Pilot-in-Command

When acting as pilot-in-command, a sport pilot must have a current pilot certificate, medical eligibility (valid and current driver's license or medical certificate), logbook endorsement for category/class and photo ID in his/her physical possession or readily accessible in the aircraft. Pilots are responsible for determining whether they are fit to fly for a particular flight.

A pilot must present his/her pilot certificate for inspection upon request of any FAA, NTSB, TSA, or federal, state, or local law enforcement officer.

The pilot-in-command (PIC) of an aircraft is directly responsible, and is the final authority, for the safety and operation of that aircraft. Should an emergency require immediate action, the PIC may deviate from 14 CFR Part 91 to the extent necessary in the interest of safety. Upon request, a written report of any deviation from the rules must be sent to the Administrator.

For operating at an airport with a control tower, an Air Traffic Control (ATC) clearance is authorization for an aircraft to proceed under specified traffic conditions within that tower's controlled airspace. When an ATC clearance has been obtained, no pilot-in-command may deviate from that clearance, except in an emergency, unless he/she obtains an amended clearance. If a pilot does deviate from a clearance or ATC instruction during an emergency, he/she must notify ATC of the deviation as soon as possible. If, in an emergency, a pilot is given priority over other aircraft by ATC, he/she may be requested to submit a detailed report even though no deviation from a rule occurred. The requested report shall be submitted within 48 hours to the chief of the ATC facility which granted the priority. If a pilot receives a clearance that would cause a deviation from a rule, he/she should query the controller and request that the clearance be amended. Sport pilots wanting to fly in airspace where there is a control tower will need a logbook endorsement to do so. Many Sport Pilots will not fly in tower-controlled airspace and so will not be required to talk with ATC.

ALL

2078. Who is responsible for determining whether a pilot is fit to fly for a particular flight, even though the pilot holds a current and valid U.S. driver's license?

A—The FAA.
B—The pilot.
C—The medical examiner.

The pilot is responsible for determining whether he/she is fit to fly for a particular flight. (PLT443) — FAA-H-8083-25

ALL

2129. Each person who holds a pilot certificate, a U.S. driver's license, or a medical certificate shall present it for inspection upon the request of the Administrator, the National Transportation Safety Board, or any

A—authorized representative of the Department of Transportation.
B—authorized representative of the Department of State.
C—federal, state, or local law enforcement officer.

Each person who holds a pilot or medical certificate shall present it for inspection upon the request of the FAA Administrator, an NTSB representative, or any Federal, State, or local law enforcement officer. (PLT399) — 14 CFR §61.3

ALL

2141. If an in-flight emergency requires immediate action, the pilot in command may

A—deviate from the FAR's to the extent required to meet the emergency, but must submit a written report to the Administrator within 24 hours.
B—deviate from any rule of 14 CFR 91 to the extent required to meet that emergency.
C—not deviate from any rule of 14 CFR 91 unless prior to the deviation, the Administrator grants approval.

If an emergency requires immediate action, the pilot-in-command may deviate from the operating rules of Part 91 to the extent necessary to meet that emergency. No report of such deviation is required unless the FAA requests one. (PLT403) — 14 CFR §91.3

Answers

2078 [B]	2129 [C]	2141 [B]

ALL

2142. When must a pilot who deviates from a regulation during an emergency send a written report of that deviation to the Administrator?

A—Within 7 days.
B—Within 10 days.
C—Upon request.

Each pilot-in-command who deviates from a rule in an emergency shall, upon request, send a written report of that deviation to the Administrator. (PLT403) — 14 CFR §91.3

ALL

2188. When would a pilot be required to submit a detailed report of an emergency which caused the pilot to deviate from an ATC clearance?

A—When requested by ATC.
B—Immediately.
C—Within 7 days.

Each pilot-in-command who deviated from an ATC clearance during an emergency must submit a detailed report when requested by ATC. (PLT403) — 14 CFR §91.123

LTA

2308. The person directly responsible for the pre-launch briefing of passengers for a flight is the

A—safety officer.
B—pilot in command.
C—ground crewmember.

The pilot-in-command is responsible for briefing crewmembers and occupants in all areas of the flight, including inflation, tether, inflight, landing, emergency, and recovery procedures. (PLT444) — 14 CFR §91.3

Alcohol and Drugs

As little as one ounce of liquor, one bottle of beer, or four ounces of wine can impair flying skills. No person may act as a crewmember on an aircraft under the following conditions:

1. Within 8 hours after the consumption of any alcoholic beverage;
2. While under the influence of alcohol (.04 percent by weight or more alcohol in the blood);
3. While using any drug that affects his/her faculties in any way contrary to safety.

Except in an emergency, no pilot of an aircraft may allow a person who appears to be intoxicated or under the influence of drugs (except a medical patient under proper care) to be carried in that aircraft.

A conviction for the violation of any law relating to drugs or alcohol is grounds for: (1) Denial of an application for any certificate, rating, or authorization issued under Part 61 for a period of up to 1 year after the date of final conviction; or (2) Suspension or revocation of any certificate, rating, or authorization issued under Part 61.

Pilots shall provide a written report of each alcohol- or drug-related motor vehicle action to the FAA, Civil Aviation Security Division (AMC-700) not later than 60 days after the motor vehicle action.

Answers

2142 [C] 2188 [A] 2308 [B]

ALL
2087. Which is true regarding the presence of alcohol within the human body?

A—A small amount of alcohol increases vision acuity.
B—An increase in altitude decreases the adverse effect of alcohol.
C—Judgment and decision-making abilities can be adversely affected by even small amounts of alcohol.

As little as one ounce of liquor, one bottle of beer, or four ounces of wine can impair flying skills. (PLT205) — AIM ¶8-1-1

Answer (A) is incorrect because all mental and physical activities will be decreased with even small amounts of alcohol in the bloodstream. Answer (B) is incorrect because the adverse effects of alcohol are increased as altitude is increased.

ALL
2139. No person may act or attempt to act as a crewmember of a civil aircraft with

A—.08 percent by weight or more alcohol in the blood.
B—.4 percent by weight or more alcohol in the blood.
C—.04 percent by weight or more alcohol in the blood.

No person may act, or attempt to act, as a crewmember of a civil aircraft while having .04 percent or more, by weight, alcohol in the blood. (PLT463) — 14 CFR §91.17

ALL
2126. A pilot convicted of operating an aircraft under the influence of alcohol, or using drugs that affect the person's faculties, is grounds for a

A—denial of an application for an FAA certificate, rating, or authorization issued under 14 CFR part 61.
B—written notification to the FAA Civil Aeromedical Institute (CAMI) within 60 days after the conviction.
C—written report to be filed with the FAA Civil Aviation Security Division (AMC-700) not later than 60 day after the conviction.

A conviction for the violation of any Federal or State statute relating to drugs or alcohol is grounds for: (1) Denial of an application for any certificate, rating, or authorization issued under Part 61 for a period of up to 1 year after the date of final conviction; or (2) Suspension or revocation of any certificate, rating, or authorization issued under Part 61. (PLT463) — 14 CFR §61.15

Answer (B) is incorrect because the report must be filed with AMC-700. Answer (C) is incorrect because this applies to automobile convictions, not aircraft convictions.

ALL
2127. A pilot convicted for the violation of any Federal or State statute relating to the process, manufacture, transportation, distribution, or sale of narcotic drugs is grounds for

A—a written report to be filed with the FAA Civil Aviation Security Division (AMC-700) not later than 60 days after the conviction.
B—notification of this conviction to the FAA Civil Aeromedical Institute (CAMI) within 60 days after the conviction.
C—suspension or revocation of any certificate, rating, or authorization issued under 14 CFR part 61.

A conviction for the violation of any Federal or State statute relating to the growing, processing, manufacture, sale, disposition, possession, transportation, or importation of narcotic drugs, marijuana, or depressant or stimulant drugs or substances is grounds for: (1) Denial of an application for any certificate, rating, or authorization for a period of up to 1 year after the date of final conviction; or (2) Suspension or revocation of any certificate, rating, or authorization. (PLT463) — 14 CFR §61.15

Answers (A) and (B) are incorrect because a written report to AMC-700 is only necessary for motor vehicle actions.

ALL
2128. A pilot convicted of operating a motor vehicle while either intoxicated by, impaired by, or under the influence of alcohol or a drug is required to provide a

A—written report to the FAA Civil Aeromedical Institute (CAMI) within 60 days after the motor vehicle action.
B—written report to the FAA Civil Aviation Security Division (AMC-700) not later than 60 days after the conviction.
C—notification of the conviction to an FAA Aviation Medical Examiner (AME) not later than 60 days after the motor vehicle action.

Each person holding a pilot certificate shall provide a written report of each motor vehicle action to the FAA, Civil Aviation Security Division (AMC-700), not later than 60 days after the motor vehicle action. (PLT463) — 14 CFR §61.15

Answers

2087 [C]	2139 [C]	2126 [A]	2127 [C]	2128 [B]

Formation Flight and Dropping Objects

Flying so close to another aircraft as to create a collision hazard is prohibited. If the intent is to fly formation, prior arrangement with the pilot-in-command of each aircraft is required. In any case, no person may operate an aircraft carrying passengers for hire in formation flight.

The PIC of an aircraft may not allow any object to be dropped while in flight unless reasonable precautions are taken to avoid injury or damage to persons or property on the surface.

ALL

2133. No person may operate an aircraft in formation flight

A—over a densely populated area.
B—in Class D airspace.
C—except by prior arrangement with the pilot in command of each aircraft.

No person may operate an aircraft in formation flight except by arrangement with the pilot-in-command of each aircraft in the formation. The only restriction to formation flight is when carrying passengers for hire. (PLT431) — 14 CFR §91.111

Right-of-Way Rules

When weather conditions permit, vigilance must be maintained so as to see and avoid other aircraft.

An aircraft in distress has the right-of-way over all other air traffic. When aircraft of the same category are converging at approximately the same altitude (except head-on, or nearly so), the aircraft on the other's right has the right-of-way. *See* Figure 1-2.

If the aircraft are of different categories, the following applies:

1. A balloon has the right-of-way over any other category of aircraft;
2. A glider has the right-of-way over an airship, powered parachute, weight-shift control aircraft, airplane, or rotorcraft; and
3. An airship has the right-of-way over a powered parachute, weight-shift control aircraft, airplane or rotorcraft.

An aircraft towing or refueling other aircraft has the right-of-way over all other engine driven aircraft.

When aircraft are approaching each other head-on, or nearly so, each pilot of each aircraft (regardless of category), shall alter course to the right similar to when driving a car (Figure 1-3).

An aircraft being overtaken has the right-of-way. The overtaking aircraft shall alter course to the right to pass well clear. This is different from the procedure when driving a car because the pilot overtaking the other aircraft typically sits on the left and can clearly see the aircraft they are overtaking. (*See* Figure 1-4).

When two or more aircraft are approaching an airport for landing, the aircraft at the lower altitude has the right-of-way, but it shall not take advantage of this rule to cut in front of, or overtake another aircraft.

Answers
2133 [C]

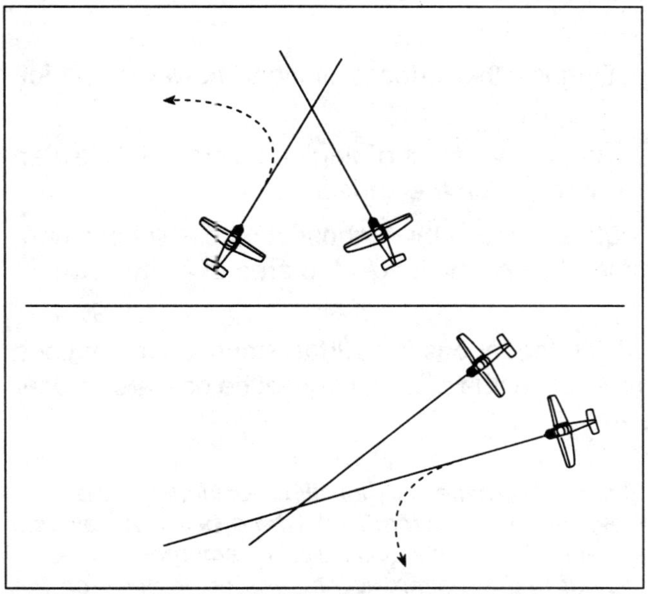

Figure 1-2. Aircraft on converging courses: Aircraft on the right has right-of-way— aircraft on the left must yield.

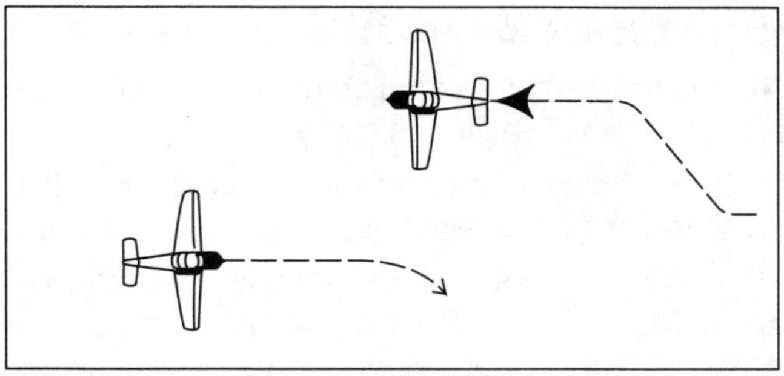

Figure 1-3. Aircraft approaching head-on

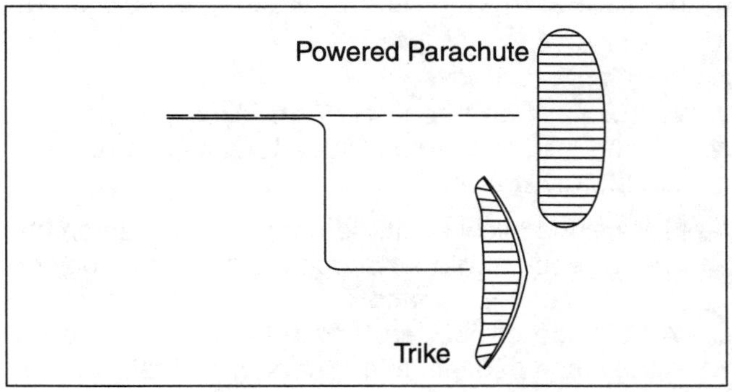

Figure 1-4. One aircraft overtaking another

ALL
2134. Which aircraft has the right-of-way over all other air traffic?

A—A balloon.
B—An aircraft in distress.
C—An aircraft on final approach to land.

An aircraft in distress has the right-of-way over all other air traffic. (PLT414) — 14 CFR §91.113

ALL
2135. What action is required when two aircraft of the same category converge, but not head-on?

A—The faster aircraft shall give way.
B—The aircraft on the left shall give way.
C—Each aircraft shall give way to the right.

When two aircraft of the same "right-of-way" category converge at approximately the same altitude, the aircraft to the other's right has the right-of-way. (PLT414) — 14 CFR §91.113

Answer (A) is incorrect because speed has nothing to do with right-of-way. Answer (C) is incorrect because "each aircraft giving way to the right" is the rule when approaching head-on.

Answers
2134 [B] 2135 [B]

Minimum Safe Altitudes

No minimum altitude applies during takeoff or landing. During other phases of flight, however, the following minimum altitudes apply:

Anywhere—The pilot must maintain an altitude which, in the event of engine failure, will allow an emergency landing without undue hazard to persons or property on the surface.

Over congested areas—An altitude of at least 1,000 feet above the highest obstacle within a horizontal radius of 2,000 feet of the aircraft must be maintained over any congested area of a city, town, or settlement or over any open-air assembly of people.

Over other than congested areas—An altitude of 500 feet above the surface must be maintained except over open water or sparsely populated areas. In that case, the aircraft may not be operated closer than 500 feet to any person, vessel, vehicle or structure.

ALL

2136. Except when necessary for takeoff or landing, what is the minimum safe altitude for a pilot to operate an aircraft anywhere?

A—An altitude allowing, if a power unit fails, an emergency landing without undue hazard to persons or property on the surface.
B—An altitude of 500 feet above the surface and no closer than 500 feet to any person, vessel, vehicle, or structure.
C—An altitude of 500 feet above the highest obstacle within a horizontal radius of 1,000 feet.

Except when necessary for takeoff or landing, no person may operate an aircraft anywhere below an altitude allowing, if a power unit fails, an emergency landing without undue hazard to persons or property on the surface. (PLT430) — 14 CFR §91.119

Answers
2136 [A]

Basic VFR Weather Minimums

See Chapter 5 for details on airspace.

Rules governing flight under **Visual Flight Rules (VFR)** have been adopted to assist the pilot in meeting his/her responsibility to see and avoid other aircraft. Sport pilots are required to maintain 3 miles visibility.

In addition, when operating within the lateral boundaries of the surface area of Class B, C, D or E airspace designated for an airport, the ceiling must not be less than 1,000 feet. If the pilot intends to land, take off, or enter a traffic pattern within such airspace, the ground visibility must be at least 3 miles at that airport. If ground visibility is not reported, 3 miles flight visibility is required in the pattern.

You need to maintain this distance from clouds: 500 feet below, 1,000 feet above, and 2,000 feet horizontal. In Class G airspace, typically up to 700 or 1,200 feet AGL, you must have 3 miles visibility and be clear of clouds.

ALL
2137. During operations within Class E airspace at altitudes of less than 1,200 feet AGL, the minimum horizontal distance from clouds requirement for flight is

A—500 feet.
B—1,000 feet.
C—2,000 feet.

Minimum horizontal distance from clouds within Class C, D, or E airspace below 10,000 feet is 2,000 feet. (PLT163) — 14 CFR §91.155

ALL
2040. Sport Pilot minimum flight visibility for Class E airspace less than 10,000 feet mean sea level (MSL) is

A—2,000 feet horizontal.
B—3 statute miles.
C—3 nautical miles.

The minimum flight visibility for Class E airspace less than 10,000 feet MSL is 3 SM. (PLT163) — FAA-H-8083-25

Answer (A) is incorrect because this is the required distance from clouds, not flight visibility. Answer (C) is incorrect because flight visibility is based on statute miles, not nautical.

ALL
2186. No person may take off or land an aircraft under basic VFR at an airport that lies within Class D airspace unless the

A—flight visibility at that airport is at least 1 mile.
B—ground visibility at that airport is at least 1 mile.
C—ground visibility at that airport is at least 3 miles.

No person may takeoff or land an aircraft, or enter the traffic pattern of an airport under VFR, within Class D airspace unless ground visibility at that airport is at least 3 statute miles. (PLT467) — 14 CFR §91.155

ALL
2187. The basic VFR weather minimums for operating an aircraft within Class D airspace are

A—500-foot ceiling and 1 mile visibility.
B—1,000-foot ceiling and 3 miles visibility.
C—clear of clouds and 2 miles visibility.

No person may operate VFR within Class D airspace, beneath the ceiling when the ceiling is less than 1,000 feet. No person may takeoff or land an aircraft, or enter the traffic pattern of an airport under VFR, within Class D airspace unless ground visibility at that airport is at least 3 statute miles. (PLT467) — 14 CFR §91.155

ALL
2203. You need to maintain this distance below clouds:

A—500 feet.
B—1,000 feet.
C—2,000 feet.

Sport pilots need to maintain this distance from clouds: 500 feet below, 1,000 feet above, and 2,000 feet horizontal in all controlled airspace. (PLT163) — 14 CFR §91.155

Answer (B) is incorrect because this is the distance you must maintain above clouds. Answer (C) is incorrect because this is the distance you must maintain horizontally from clouds.

Answers

2137 [C]	2040 [B]	2186 [C]	2187 [B]	2203 [A]

Airworthiness and Light-Sport Aircraft Definitions

Each aircraft is issued an Airworthiness Certificate, which remains valid as long as the aircraft is maintained and operated as required by regulations. This Airworthiness Certificate, along with the Aircraft Registration Certificate, operating limitations, and weight and balance must be aboard the aircraft during flight. Remember: **A**irworthiness • **R**egistration • **O**perating Limitations • **W**eight and balance

The aircraft's operating limitations and weight and balance may be found in the airplane flight manual, approved manual material, markings, and placards, or any combination thereof.

Summarized below are important points in understanding the definition of light-sport aircraft and the privileges associated with the Sport Pilot certificate:

- Two-place maximum (pilot and one passenger)
- Maximum gross takeoff weight 1,320 lbs (599 kg); 1,430 lbs for seaplanes
- Lighter-than-air light-sport aircraft maximum gross weight 660 lbs (300 kg)
- Maximum stall speed 51.8 mph (45 knots)
- Maximum speed in level flight with maximum continuous power 138 mph (120 knots)
- All LSA aircraft logbook signoffs are based on a specific make and model, allowing you to fly a similar "set" of performance and handling aircraft. Aircraft are organized by the FAA from large "Categories" of aircraft to specific "Sets" in the following order:

  Category and Class Light-Sport Aircraft — fixed wing/airplane category (land and sea classes), weight-shift control/trike category (land and sea classes), powered parachute category (land and sea classes), glider, rotorcraft category (gyroplane class), lighter-than-air category (balloon and airship classes).

  Sets are a more detailed breakdown of similar performance and handling.

  The 8 Airplane sets are:

  Less than 87 KCAS V_H
  1. Tricycle gear
  2. Tailwheel
  3. Float equipped
  4. Ski equipped

  Greater than 87 KCAS V_H
  5. Tricycle gear
  6. Tailwheel
  7. Float equipped
  8. Ski equipped

  The 3 Weight-Shift Control sets are:
  1. Tricycle gear
  2. Float equipped
  3. Ski equipped

  The 4 Powered Parachute sets are:

  Square wing
  1. Land
  2. Sea

  Elliptical Wing
  3. Land
  4. Sea

  Make and Model — manufacturer specific that define/specify a set.

  Type — E-LSA or S-LSA.
- Single, non-turbine engine only, includes rotary or diesel engines
- Fixed or ground-adjustable propeller
- Unpressurized cabin with fixed landing gear
- Seaplanes with landing gear that can be configured so the wheels rotate for amphibious operation
- Will have FAA registration N number

- U.S. or foreign manufacture of light-sport aircraft is authorized
- (S-LSA) Special Light-Sport aircraft manufactured and sold ready-to-fly under a new certification without 14 CFR Part 23 compliance, but must meet ASTM consensus standards. Aircraft under this certification may be used for sport and recreation, flight training, and aircraft rental.
- (E-LSA) Experimental Light-Sport Aircraft is a kit built from an S-LSA design. May be used for sport and recreation and flight instruction for the owner of the aircraft.
- (E-LSA) Experimental Light-Sport Aircraft which was operated as an ultralight trainer or overweight single-place ultralight (fat ultralight). Aircraft must be transitioned to E-LSA category no later than January 31, 2008. This existing fleet of ultralight trainers can be used for training until January 31, 2010.
- Aircraft with a standard airworthiness certificate (vintage production aircraft) that meet LSA specifications may be flown by sport pilots. However, that airworthiness certification category will not be changed to a light-sport aircraft. Maintenance must therefore be done by an A&P as it has in the past.
- Holders of a Sport Pilot Certificate may fly an aircraft with a standard airworthiness certificate if it meets the definition of a light-sport aircraft.
- May be operated at night if the aircraft is equipped per 14 CFR §91.209 and the pilot holds at least a Private Pilot certificate and a minimum of a third-class medical.
- E-LSA Experimental amateur-built, in which the builder completes 51% of the building of the aircraft.

ALL

2190. How long does the Airworthiness Certificate of an aircraft remain valid?

A—As long as the aircraft has a current Registration Certificate.

B—Indefinitely, unless the aircraft suffers major damage.

C—As long as the aircraft is maintained and operated as required by Federal Aviation Regulations.

Unless sooner surrendered, suspended, revoked, or a termination date is otherwise established by the Administrator, Standard Airworthiness Certificates and Airworthiness Certificates issued for restricted or limited category aircraft are effective as long as the maintenance, preventive maintenance, and alterations are performed in accordance with Parts 43 and 91 and the aircraft are registered in the United States. (PLT377) — 14 CFR §21.181

ALL

2140. In addition to a valid Airworthiness Certificate, what documents or records must be aboard an aircraft during flight?

A—Aircraft engine and airframe logbooks, and owner's manual.

B—Radio operator's permit, and repair and alteration forms.

C—Operating limitations and Registration Certificate.

No person may operate an LSA unless it has within it:

1. *An appropriate and current Airworthiness Certificate, displayed at the cabin or cockpit entrance so that it is legible to passengers or crew.*
2. *A Registration Certificate issued to its owner.*
3. *An approved flight manual (Pilot's Operating Handbook), weight and balance, markings and placards, or any combination thereof, which show the operating limitations of the aircraft.*

(PLT400) — 14 CFR §91.203

Answers

2190 [C] 2140 [C]

Maintenance and Inspections

The owner or operator of an aircraft is primarily responsible for maintaining that aircraft in an airworthy condition, including compliance with safety directives for LSA or airworthiness directives for standard category aircraft. He/she is also responsible for all records of maintenance, repairs, and alterations. (The pilot-in-command is responsible for determining that the aircraft is in an airworthy condition prior to flight.) The airworthiness of an aircraft can be determined by a preflight inspection and a review of the maintenance records.

The holder of a pilot certificate is allowed (within certain limits) to perform preventive maintenance on any aircraft owned and operated by that pilot. Preventive maintenance is limited to tasks such as replacing defective safety wire, servicing landing gear wheel bearings, replacing safety belts, and other tasks listed in 14 CFR Part 43, Appendix A. After preventive maintenance has been performed on an aircraft, the signature, certificate type and certificate number of the person approving the aircraft for return to service, and a description of the work, must be entered in the aircraft maintenance records.

An aircraft may not be operated unless, within the preceding 12 calendar months, it has had an annual inspection and has been approved for return to service. This will be indicated by the appropriate notation in the aircraft maintenance records. To determine the expiration date of the last annual inspection, refer to the aircraft maintenance records. For example, if the aircraft's last annual inspection was performed on July 12, 2004, the next annual inspection will be due no later than midnight, July 31, 2005.

If an aircraft is used to carry passengers for hire or used for flight instruction for hire, it must have, in addition to the annual inspection, an inspection each 100 hours of flight time. The 100-hour limitation may be exceeded by not more than 10 hours if necessary to reach a place at which the inspection can be done. The excess time, however, is not included in computing the next 100 hours of time in service. For example: a 100-hour inspection was due at 302.5 hours on the tachometer. The 100-hour inspection was actually done at 309.5 hours. The next 100-hour inspection is due at 402.5 hours, (100 hours after the previous inspection was due, not after the time the inspection was actually completed).

The transponder cannot be operated unless within the preceding 24 calendar months, it has been inspected and found satisfactory.

If an aircraft has been maintained, rebuilt, or altered in any manner that may have appreciably changed its flight characteristics or operation in flight, no passengers may be carried until it has been flight tested by an appropriately-rated pilot and approved for return to service.

Light-Sport Repairman Certificates

Two additional repairman certificates are available for working on light-sport aircraft: the maintenance and inspection ratings.

The **maintenance rating** allows repairmen to do maintenance, preventative maintenance, and alterations to E-LSA and S-LSA. The course is 120 hours (airplane category); 104 hours (weight-shift or powered parachute); 80 hours (glider or lighter-than-air). An A&P certificate counts towards these hour requirements for Light-Sport Aircraft qualifications, but additional training on specific category of aircraft will be required.

The **inspection rating** allows E-LSA repairmen/owners the ability to perform the annual inspection on their own aircraft. This course is 16 hours on the specific aircraft category and class. The 100-hour inspection required for training and hire for all E-LSA and S-LSA must be done by an appropriately-rated A&P mechanic or an LSA repairman with a maintenance rating.

Light-Sport Aircraft FAA Designated Airworthiness Representatives (DAR) must attend a three-day FAA course to inspect and issue experimental (E-LSA) and special (S-LSA) airworthiness certificates.

Special Light-Sport Airworthiness Certificated Aircraft (S-LSA) Maintenance

A certificated pilot (Sport Pilot rating or higher) can perform preventive maintenance as specified by the manufacturer (typically similar to 14 CFR Part 43 Appendix A) and return to service on his own aircraft with no sport pilot repairman certificate. However, the annual and 100-hour condition inspection must be completed by:

- An appropriately rated A&P mechanic;
- An appropriately rated repair station; or
- A light-sport repairman with a maintenance rating.

Experimental Light-Sport Airworthiness Certificated Aircraft (E-LSA) Maintenance

No rating is required to perform maintenance on (E-LSA).

The annual condition inspection can be completed by:

- An appropriately-rated A&P mechanic;
- An appropriately-rated repair station;
- A light-sport repairman with a maintenance rating; or
- A light-sport repairman with an inspection rating (16 hour course) on their own aircraft.

ALL

2143. Who is primarily responsible for maintaining an aircraft in an airworthy condition?

A—The lead mechanic responsible for that aircraft.
B—Pilot in command or operator.
C—Owner or operator of the aircraft.

The owner or operator of an aircraft is primarily responsible for maintaining that aircraft in an airworthy condition. (PLT374) — 14 CFR §91.403

ALL

2144. Who is responsible for determining if an aircraft is in condition for safe flight?

A—A certificated aircraft mechanic.
B—The pilot in command.
C—The owner or operator.

The pilot-in-command of an aircraft is responsible for determining whether that aircraft is in condition for safe flight. The pilot shall discontinue the flight when unairworthy mechanical, electrical or structural conditions occur. (PLT444) — 14 CFR §91.7

ALL

2033-1. During the preflight inspection who is responsible for determining the aircraft is safe for flight?

A—The pilot in command.
B—The owner or operator.
C—The certificated mechanic who performed the annual inspection.

The pilot-in-command of an aircraft is responsible for determining whether that aircraft is in condition for safe flight. (PLT444) — FAA-H-8083-25

Answers

2143 [C] 2144 [B] 2033-1 [A]

Seatbelts

All required flight crewmembers must remain in their seats with seatbelts secured during the entire flight unless absent in connection with duties or physiological needs. When shoulder harnesses are installed they must be used during takeoffs and landings.

Prior to takeoff, the pilot-in-command must ensure that each person on board has been briefed on the use of seatbelts. In addition, he/she must ensure that the passengers are notified to fasten their seatbelts during taxi, takeoffs, and landings.

ALL

2145. Safety belts are required to be properly secured about which persons in an aircraft and when?

A—Pilots only, during takeoffs and landings.
B—Passengers, during taxi, takeoffs, and landings.
C—Each person on board the aircraft during the entire flight.

During taxi, takeoff and landing, each person on board the aircraft must occupy a seat or berth with a safety belt and shoulder harness, properly secured if installed. However, a person who has not reached his/her second birthday may be held by an adult who is occupying a seat or a berth, and a person on board for the purpose of engaging in sport parachuting may use the floor of the aircraft as a seat. (PLT465) — 14 CFR §91.107

ALL

2146. Which best describes the flight conditions under which pilots are specifically required to keep their safety belts and shoulder harnesses fastened?

A—Safety belts during takeoff and landing; shoulder harnesses during takeoff and landing.
B—Safety belts during takeoff and landing; shoulder harnesses during takeoff and landing and while en route.
C—Safety belts during takeoff and landing and while en route; shoulder harnesses during takeoff and landing.

During takeoff and landing, and while en route, each required flight crewmember shall keep his/her seatbelt fastened while at his/her station. During takeoff and landing this includes shoulder harness (if installed) unless it interferes with required duties. (PLT464) — 14 CFR §91.105

ALL

2232. The pilot in command is responsible for ensuring that each person on board applicable U.S. registered aircraft is briefed and instructed on how and when to

A—fasten and unfasten their seat belt and shoulder harness.
B—adjust their seat.
C—operate the fire extinguisher.

Unless otherwise authorized by the FAA, no pilot may takeoff in a civil aircraft unless the pilot-in-command ensures that each person on board is briefed on how to fasten and unfasten that person's seatbelt, and that each person has been notified to fasten the seatbelt during taxi, takeoff and landing. (PLT441) — 14 CFR §91.107

Answers

2145 [B] 2146 [C] 2232 [A]

Accident Reporting Requirements

The National Transportation Security Board (NTSB) Part 830 contains rules pertaining to notification and reporting of aircraft accidents and incidents. It also addresses preservation of aircraft wreckage, mail, cargo, and records.

The term **accident** means "an occurrence in which any person suffers death or **serious injury**, or in which the aircraft receives **substantial damage**."

Serious injury means any injury which: (1) requires hospitalization for more than 48 hours, commencing within 7 days from the date the injury was received; (2) results in a fracture of any bone (except simple fractures of fingers, toes, or nose); (3) causes severe hemorrhages, nerve, muscle, or tendon damage; (4) involves any internal organ; or (5) involves second or third degree burns, or any burns affecting more than 5 percent of the body surface.

Substantial damage means damage or failure which adversely affects the structural strength, performance, or flight characteristics of the aircraft, and which would normally require major repair or replacement of the affected component. The following is **not** considered "substantial damage": engine failure, bent fairings or cowling, dented skin, small punctured holes in the skin or fabric, ground damage to rotor or propeller blades, and damage to landing gear, wheels, tires, flaps, engine accessories, brakes, or wingtips.

The term "incident" means "an occurrence other than an **accident**, which affects or could affect the safety of operations."

Immediate notification of the NTSB is required when an aircraft accident occurs, and any of a specified list of incidents, which include:

1. Inability of a flight crewmember to perform his/her duties due to illness or injury;
2. Infight fire;
3. An aircraft is overdue and believed to have been involved in an accident;
4. Flight control system malfunction or failure; or
5. When aircraft collide in flight or damage of more than $25,000 occurs to property other than the aircraft.

The operator of an aircraft involved in an accident or incident which requires notification of the NTSB is responsible for preserving the wreckage, mail, or cargo until the NTSB takes custody. These items may be moved to protect the wreckage from further damage.

The operator of an aircraft involved in an accident is required to file an accident report within 10 days. A report of an incident must be reported only upon request.

ALL

2148. How many days after an accident is a report required to be filed with the nearest NTSB field office?

A—2.
B—7.
C—10.

The operator of an aircraft shall file a report on Board Form 6120.1 or 6120.2 within 10 days of an accident. (PLT366) — NTSB §830.15

ALL

2150. The operator of an aircraft that has been involved in an accident is required to file an accident report within how many days?

A—5.
B—7.
C—10.

The operator of an aircraft shall file a report on Board Form 6120.1 or 6120.2 within 10 days of an accident. (PLT366) — NTSB §830.15

ALL

2189. While taxiing on the parking ramp, the landing gear, wheel, and tire are damaged by striking ground equipment. What action would be required to comply with NTSB Part 830?

A—A report must be filed with the nearest FAA field office within 7 days.
B—An immediate notification must be filed by the operator of the aircraft with the nearest NTSB field office.
C—No notification or report is required.

This unfortunate event would not require notification or a report with the NTSB, as it is not an accident or considered "substantial damage" per the regulations. (PLT366) — NTSB §830.2

Answer (A) is incorrect because a report must be filed within 7 days for certain accidents or injuries, of which the taxiing incident described here does not qualify. Answer (B) is incorrect because an immediate notification must be filed only when an aircraft accident occurs, of which the taxiing incident described here does not qualify.

ALL

2149. The operator of an aircraft that has been involved in an incident is required to submit a report to the nearest field office of the NTSB

A—within 7 days.
B—within 10 days.
C—only if requested to do so.

A report on an incident for which notification is required shall be filed only as requested by an authorized representative of the Board. (PLT366) — NTSB §830.15

ALL

2151. May aircraft wreckage be moved prior to the time the NTSB takes custody?

A—Yes, but only if moved by a federal, state, or local law enforcement officer.
B—Yes, but only to protect the wreckage from further damage.
C—No, it may not be moved under any circumstances.

Prior to the time the Board and its authorized representative takes custody of aircraft wreckage, it may not be disturbed or moved except to remove persons injured or trapped, to protect the wreckage from further damage, or to protect the public from injury. (PLT366) — NTSB §830.10

ALL

2152. If an aircraft is involved in an accident which results in substantial damage to the aircraft, the nearest NTSB field office should be notified

A—immediately.
B—within 48 hours.
C—within 7 days.

The operator of an aircraft shall immediately, and by the most expeditious means available, notify the nearest NTSB field office when an aircraft accident occurs. (PLT366) — NTSB §830.5

Answers

2148 [C]	2150 [C]	2189 [C]	2149 [C]	2151 [B]	2152 [A]

ALL

2211. Which publication covers the procedures required for aircraft accident and incident reporting responsibilities for pilots?

A—FAR Part 61.
B—FAR Part 91.
C—NTSB Part 830.

NTSB Part 830 contains regulations pertaining to notification and reporting of aircraft accidents or incidents and overdue aircraft, and preservation of aircraft wreckage, mail, cargo, and records. (PLT366) — NTSB Part 830

Answer (A) is incorrect because 14 CFR Part 61 contains regulations on certification of pilots. Answer (B) is incorrect because 14 CFR Part 91 contains regulations on general operating and flight rules.

ALL

2212. Notification to the NTSB is required when there has been substantial damage

A—which requires repairs to landing gear.
B—to an engine caused by engine failure in flight.
C—which adversely affects structural strength or flight characteristics.

The operator of an aircraft shall immediately, and by the most expeditious means available, notify the nearest NTSB field office when an aircraft accident occurs. An aircraft accident is an occurrence associated with the operation of an aircraft which takes place between the time any person boards the aircraft with the intention of flight and all such persons have disembarked, and in which any person suffers death or serious injury, or in which the aircraft receives substantial damage which adversely affects the structural strength, performance, or flight characteristics of the aircraft. (PLT366) — 49 CFR §830.2, §830.5

ALL

2213. What period of time must a person be hospitalized before an injury may be defined by the NTSB as a "serious injury"?

A—10 days, with no other extenuating circumstances.
B—48 hours; commencing within 7 days after date of the injury.
C—72 hours; commencing within 10 days after date of the injury.

"Serious injury" means any injury requiring hospitalization for more than 48 hours, commencing within 7 days from the date the injury was received. (PLT395) — 49 CFR §830.2

ALL

2214. Which incident would require that the nearest NTSB field office be notified immediately?

A—In-flight fire.
B—Ground fire resulting in fire equipment dispatch.
C—Fire of the primary aircraft while in a hangar which results in damage to other property of more than $25,000.

The operator of an aircraft shall immediately and by the most expeditious means available notify the nearest NTSB field office of any infight fire. (PLT416) — 49 CFR §830.5

Answers (B) and (C) are incorrect because the regulation specifies "infight" fires only.

Answers

2211 [C]	2212 [C]	2213 [B]	2214 [A]

FAA Publications

The FAA publication that provides the aviation community with basic flight information and Air Traffic Control procedures for use in the National Airspace System of the United States is the ***Aeronautical Information Manual*** **(AIM)**. It is the official guide to basic flight information and ATC procedures.

Advisory Circulars (AC) are issued by the FAA to inform the aviation community in a systematic way of non-regulatory material of interest. In many cases, they are the result of a need to fully explain a particular subject (wake turbulence, for example). They are issued in a numbered-subject system corresponding to the subject areas of the Federal Aviation Regulations. Advisory Circulars (some free, others at cost) may be obtained by ordering from the Government Printing Office (GPO). Appendix II of the Advisory Circular Checklist contains the Circular Numbering System wherein AC numbers relate to FAR subchapter titles and correspond to the Parts and/or sections of the regulations. The four to remember are: 20–Aircraft; 60–Airmen; 70–Airspace; and 90–Air Traffic and General Operating Rules.

Airworthiness Directives (ADs) for standard category aircraft identify unsafe aircraft conditions and prescribe regulatory actions (such as inspections or modifications) or limitations under which the affected aircraft may continue to be operated and are mandatory. Compliance with an applicable Airworthiness Directive must be entered in the appropriate aircraft maintenance records. The owner or operator is responsible for ensuring ADs are complied with. Pilots may operate an aircraft not in compliance with an AD if the AD allows for this.

Safety Directives (SD) for light-sport aircraft identify unsafe aircraft conditions and prescribe regulatory actions (such as inspections or modifications) or limitations under which the affected aircraft may continue to be operated and are mandatory. Applicable Safety Directive compliance must be entered in the appropriate aircraft maintenance records. The owner or operator is responsible for ensuring SD compliance. Pilots may operate an aircraft not in compliance with an SD if the SD allows for this.

Notices to Airmen (NOTAMs) provide the most current information available. They provide time-critical information on airports and changes that affect the national airspace system and are of concern to instrument flight rule (IFR) operations. NOTAM information is classified into three categories. These are NOTAM-D or distant, NOTAM-L or local, and flight data center (FDC) NOTAMs. NOTAM-Ds are attached to hourly weather reports and are available at flight service stations (AFSS/FSS). NOTAM-Ls include items of a local nature, such as taxiway closures or construction near a runway. These NOTAMs are maintained at the FSS nearest the airport affected. NOTAM-Ls must be requested from the FSS for which the NOTAM was issued. FDC NOTAMs are issued by the National Flight Data Center and contain regulatory information, such as temporary flight restrictions or an amendment to instrument approach procedures. The NOTAM-Ds and FDC NOTAMs are contained in the Notices to Airmen publication, which is issued every 28 days. Prior to any flight, pilots should check for any NOTAMs that could affect their intended flight.

An **FDC NOTAM** will be issued to designate a temporary flight restriction (TFR). The NOTAM will begin with the phrase "FLIGHT RESTRICTIONS" followed by the location of the temporary restriction, effective time period, area defined in statute miles, and altitudes affected. The NOTAM will also contain the FAA coordination facility and telephone number, the reason for the restriction, and any other information deemed appropriate. The pilot should check the NOTAMs as part of flight planning. Some of the purposes for establishing a temporary restriction are: protect persons and property in the air or on the surface from an existing or imminent hazard, provide a safe environment for the operation of disaster relief aircraft, prevent an unsafe congestion of sightseeing aircraft above an incident or event, which may generate a high degree of public interest, protect declared national disasters for humanitarian reasons in the State of Hawaii, protect the President, Vice President, or other public figures, provide a safe environment for space agency operations.

The **Airport/Facility Directory (A/FD)** provides the most comprehensive information on a given airport. It contains information on airports, heliports, and seaplane bases that are open to the public. The A/FDs are contained in seven books, which are organized by regions. The A/FDs are revised every 8 weeks. VFR Sectional Charts are revised twice a year (so the A/FD usually has more current information).

ALL

2107. The Airport/Facility Directory (A/FD) will generally have the latest information pertaining to airport elevation, runway facilities, and control tower frequencies. If there are differences, it should be used in preference to the information

A—on the sectional chart.
B—in the Pilot's Handbook of Aeronautical Knowledge.
C—in the Aeronautical Information Manual (AIM).

A/FDs are revised every 8 weeks; sectional charts are revised semiannually. If there are differences, the A/FD should be relied upon since they are more current. (PLT078) — FAA-H-8083-25

Answers (B) and (C) are incorrect because these publications do not have airport specific information.

ALL

2038. (Refer to Figure 66, area 2 and Legend 1.) For information about the parachute jumping and glider operations at Silverwood Airport, refer to

A—notes on the border of the chart.
B—the Airport/Facility Directory.
C—the Notices to Airmen (NOTAM) publication.

Tabulations of parachute jump areas in the U.S. are contained in the Airport/Facility Directory. (PLT064) — AIM ¶3-5-5

ALL

2180. FAA advisory circulars containing subject matter specifically related to Air Traffic Control and General Operations are issued under which subject number?

A—60.
B—70.
C—90.

Appendix II of the Advisory Circular Checklist contains the Circular Numbering System wherein AC numbers relate to FAR subchapter titles and correspond to the Parts and/or sections of the regulations. The four to remember are:

20—Aircraft;

60—Airmen;

70—Airspace; and

90—Air Traffic and General Operating Rules

(PLT116) — AC 00-2

ALL

2109. Unless incorporated into a regulation by reference, Advisory Circulars (ACs) are issued to inform the public of nonregulatory material

A—and are not binding.
B—but are binding.
C—and self-cancel after 1 year.

The FAA issues advisory circulars to inform the aviation public in a systematic way of nonregulatory material. Unless incorporated into a regulation by reference, the contents of an advisory circular are not binding on the public. (PLT116) — AC 00-2

ALL

2110. Some Advisory Circulars (ACs) are available free of charge while the remaining ACs must be purchased. All aviation safety ACs may be obtained by following the procedures in the AC Checklist (AC 00-2) or by

A—referring to the FAA internet home page and following the links to ACs.
B—contacting the local airport Fixed Base Operator and requesting the desired AC.
C—reading the ACs in the Aeronautical Information Manual (AIM).

A number of FAA organizations have made their ACs available on the Internet. These ACs may be found through the FAA home page at: www.faa.gov located under Regulatory/Advisory, select Advisory Circulars. (PLT116) — AC 00-2

Answers (B) and (C) are incorrect because FBOs are not responsible for distributing ACs, nor are they included as part of the AIM.

Answers

2107 [A] 2038 [B] 2180 [C] 2109 [A] 2110 [A]

ALL
2117. The Federal Aviation Administration publication that provides the aviation community with basic flight information and Air Traffic Control procedures for use in the National Airspace System of the United States is the

A—Aeronautical Information Manual (AIM).
B—Airport/Facility Directory (A/FD).
C—Advisory Circular Checklist (AC 00-2).

The Aeronautical Information Manual (AIM) *is the official guide to basic flight information and ATC procedures. (PLT116) — AIM*

Answer (B) is incorrect because the A/FD contains information on airports, communications, NAVAIDs, instrument landing systems, VOR receiver check points, preferred routes, Flight Service Station/Weather Service telephone numbers, Air Route Traffic Control Center (ARTCC) frequencies, part-time surface areas, and various other pertinent special notices essential to air navigation. Answer (C) is incorrect because AC 00-2 is the Advisory Circular checklist and status of other FAA publications, contains advisory circulars that are for sale as well as those distributed free-of-charge by the FAA.

ALL
2118. The most comprehensive information on a given airport is provided by

A—the Airport/Facility Directory (A/FD).
B—Notices to Airmen (NOTAMS).
C—world aeronautical (WAC) charts.

The Airport/Facility Directory (A/FD) provides the most comprehensive information on a given airport. (PLT281) — FAA-H-8083-25

Answer (B) is incorrect because NOTAMs provide time-critical information on airports and changes that affect the national airspace system. Answer (C) is incorrect because WACs provide a standard series of aeronautical charts covering land areas of the world at a size and scale convenient for navigation by moderate speed aircraft.

ALL
2119. For a complete listing of information provided in an Airport/Facility Directory (A/FD) and how the information may be decoded, refer to the

A—'Directory Legend Sample' located in the front of each A/FD.
B—Aeronautical Information Manual (AIM).
C—legend on sectional, VFR terminal area, and world aeronautical charts.

For a complete listing of information provided in an A/FD and how the information may be decoded, refer to the "Directory Legend Sample" located in the front of each A/FD. (PLT078) — FAA-H-8083-25

ALL
2120. Flight Data Center (FDC) NOTAMS are issued by the National Flight Data Center and contain regulatory information, such as

A—temporary flight restrictions.
B—markings and signs used at airports.
C—standard communication procedures at uncontrolled airports.

FDC NOTAMs are issued by the National Flight Data Center and contain regulatory information, such as temporary flight restrictions. (PLT323) — FAA-H-8083-25

ALL
2121. Time-critical information on airports and changes that affect the national airspace system are provided by

A—Notices to Airmen (NOTAMS).
B—the Airport/Facilities Directory (A/FD).
C—Advisory Circulars (ACs).

Notices to Airmen (NOTAMs) provide the most current information available. They provide time-critical information on airports and changes that affect the national airspace system. (PLT323) — FAA-H-8083-25

Answer (B) is incorrect because A/FDs are only revised every 8 weeks. Answer (C) is incorrect because the FAA issues ACs to inform the aviation public in a systematic way of nonregulatory material.

ALL
2122. NOTAM-Ls (local NOTAMS) include items of a local nature. NOTAM-Ls are maintained at each Flight Service Station (FSS) for facilities in their area only. NOTAM-L information for other FSS areas must be specifically requested from the FSS

A—that has responsibility for the airport concerned.
B—with which the pilot communicates.
C—where the flight plan is filed.

NOTAM-Ls include items of a local nature, such as taxiway closures or construction near a runway. These NOTAMs are maintained at the FSS nearest the affected airport. NOTAM-Ls must be requested from an FSS for which the NOTAM was issued. (PLT323) — FAA-H-8083-25

Answers

2117 [A]	2118 [A]	2119 [A]	2120 [A]	2121 [A]	2122 [A]

ALL

2053. One of the purposes for issuing a Temporary Flight Restriction (TFR) is to

A—announce Parachute Jump Areas.
B—protect public figures.
C—identify Airport Advisory Areas.

Temporary flight restrictions (TFR) are imposed in order to:

1. *Protect persons and property in the air or on the surface from an existing or imminent flight associated hazard;*
2. *Provide a safe environment for the operation of disaster relief aircraft;*
3. *Prevent an unsafe congestion of sightseeing aircraft above an incident;*
4. *Protect the President, Vice President, or other public figures; and,*
5. *Provide a safe environment for space agency operations.*

Pilots are expected to check appropriate NOTAMs during flight planning when conducting flight in an area where a temporary flight restriction is in effect. (PLT376) — FAA-H-8083-15, Glossary

Answer (A) is incorrect because this is done through NOTAMs. Answer (C) is incorrect because these are identified through the Sectional and A/FD.

ALL

2125. May a pilot operate an aircraft that is not in compliance with an Airworthiness Directive (AD)?

A—Yes, AD's are only voluntary.
B—Yes, if allowed by the AD.
C—Yes, under VFR conditions only.

No person may operate a product to which an airworthiness directive applies except in accordance with the requirements of that airworthiness directive. (PLT378) — 14 CFR §39.3

Answers

2053 [B] 2125 [B]

Chapter 2
Weather and Weather Services

The Global Energy Engine

Understanding how the weather starts and how it works will help you understand the dynamic environment in which you fly. Examining the global energy engine will help you predict what the weather will be like for an upcoming flight today and for the coming days.

The major source of all weather is the sun. Changes or variations of weather patterns are caused by the unequal heating of the Earth's surface.

Every physical process of weather is accompanied by or is a result of unequal heating of the Earth's surface. The heating of the Earth (and therefore the heating of the air surrounding the Earth) is unequal around the entire planet. Both north or south of the equator, one square foot of sunrays is not concentrated over one square foot of the surface, but over a larger area. This lower concentration of sunrays produces less radiation of heat over a given surface area; therefore, less atmospheric heating takes place in that area.

The unequal heating of the Earth's atmosphere creates a large air-cell circulation pattern (wind) because the warmer air has a tendency to rise (low pressure) and the colder air has a tendency to settle or descend (high pressure) and replace the rising warmer air. This unequal heating, which causes pressure variations, will also cause variations in altimeter settings between weather reporting points.

Because of the earth's tremendous rotational momentum, this large simple air-cell circulation pattern is greatly distorted by a phenomenon known as the **Coriolis force**. When the wind (which is created by high pressure trying to flow into low pressure) first begins to move at higher altitudes, the Coriolis force deflects it to the right (in the Northern Hemisphere) causing it to flow parallel to the isobars (lines of equal pressure). These deflections of the large-cell circulation pattern create general wind patterns. *See* Figure 2-1.

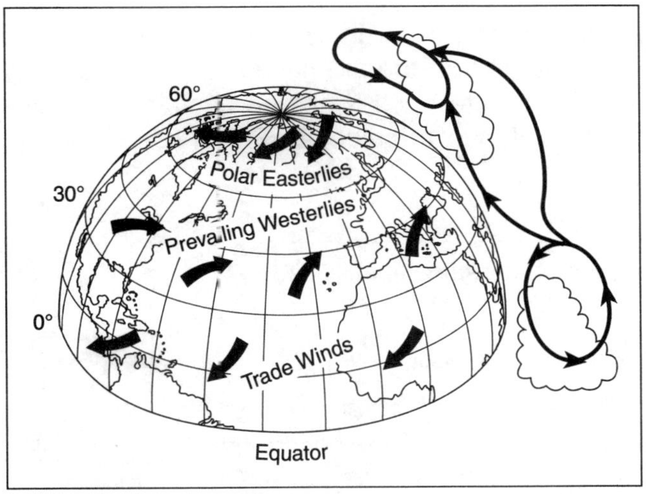

Figure 2-1. Prevailing wind systems

ALL

2171. Every physical process of weather is accompanied by or is the result of

A—a heat exchange.
B—the movement of air.
C—a pressure differential.

Every physical process of weather is accompanied by or is a result of unequal heating of the Earth's surface. (PLT510) — AC 00-6A, Chapter 2

ALL

2181. What causes variations in altimeter settings between weather reporting points?

A—Unequal heating of the Earth's surface.
B—Variation of terrain elevation.
C—Coriolis force.

All altimeter settings are corrected to sea level. Unequal heating of the Earth's surface causes pressure differences. (PLT165) — AC 00-6A, Chapter 2

Answers
2171 [A] 2181 [A]

LTA
2309. What condition does a rising barometer indicate for balloon operations?

A—Decreasing clouds and wind.
B—Chances of thunderstorms.
C—Approaching frontal activity.

Within a high-pressure system, flying conditions are generally more favorable than in low-pressure areas because there are normally fewer clouds, better visibility, calm or light winds, and less turbulence. (PLT177) — AC 00-6A, Chapter 4

LTA
2316. An airship descending through a steep temperature inversion will

A—show no change in superheat as altitude is lost.
B—show a decrease in superheat as altitude is lost.
C—become progressively lighter, thus becoming increasingly more difficult to drive down.

As the airship descends into the colder temperature of the inversion, the weight of the displaced air volume is increasing, thereby increasing buoyancy. (PLT125) — Goodyear Airship Operations Manual

Air Masses and Fronts

When a body of air comes to rest on, or moves slowly over, an extensive area having fairly uniform properties of temperature and moisture, the air takes on these properties. The area over which the air mass acquires its identifying distribution of temperature and moisture is its "source region." As this air mass moves from its source region, it tends to take on the properties of the new underlying surface. The trend toward change is called air mass modification.

A **ridge** is an elongated area of high pressure and an area of descending air. A **trough** is an elongated area of low pressure and an area of rising air. All fronts lie in troughs. A **cold front** is the leading edge of an advancing cold air mass. A **warm front** is the leading edge of an advancing warm air mass. Warm fronts move about half as fast as cold fronts. **Frontal waves** and **cyclones** (areas of low pressure) usually form on slow-moving cold fronts or stationary fronts.

The physical manifestations of a warm or cold front can be different with each front. They vary with the speed of the air mass on the move and the degree of stability of the air mass being overtaken. A stable air mass forced aloft will continue to exhibit stable characteristics, while an unstable air mass forced to ascend will continue to be characterized by cumulus clouds, turbulence, showery precipitation, and good visibility.

Frontal passage will be indicated by the following discontinuities:

1. A temperature change (the most easily recognizable discontinuity);
2. A continuous decrease in pressure followed by an increase as the front passes; and
3. A shift in the wind direction, speed, or both.

Figure 2-2 shows the symbols that would appear on a weather map.

Symbols on Surface Analysis (Surface Weather Maps)

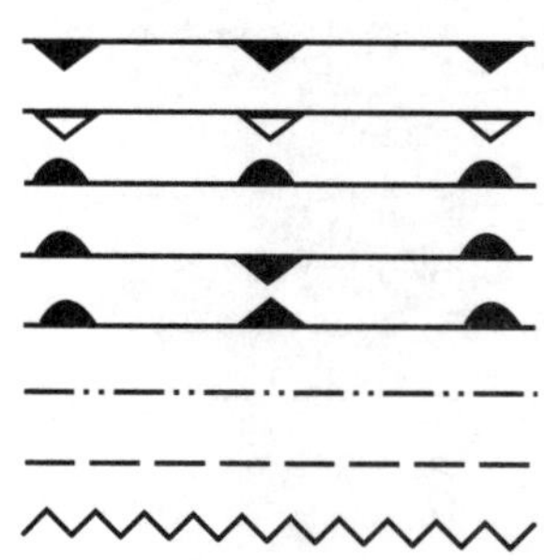

Color	Symbol	Description
Blue		Cold front
Blue		Cold front aloft
Red		Warm front
Red/Blue		Stationary front
Purple		Occluded front
Purple		Squall line
Brown		Trough
Yellow		Ridge

Figure 2-2. Weather map symbols

Answers
2309 [A] 2316 [C]

ALL
2170. Which is true with respect to a high- or low-pressure system?

A—A high-pressure area or ridge is an area of rising air.
B—A low-pressure area or trough is an area of descending air.
C—A high-pressure area or ridge is an area of descending air.

Highs and ridges are areas of descending air. (PLT517) — AC 00-6A, Chapter 4

Answer (A) is incorrect because a high or ridge is an area of descending air. Answer (B) is incorrect because a low or trough is an area of rising air.

GLI
2312. During which period is a sea breeze front most suitable for soaring flight?

A—Shortly after sunrise.
B—During the early forenoon.
C—During the afternoon.

A sea breeze begins during early afternoon and reaches a maximum in the afternoon, subsiding around dusk. (PLT511) — AC 00-6A, Chapter 16

Moisture

Air has moisture (water vapor) in it. The water vapor content of air can be expressed in two different ways. The two commonly used terms are **relative humidity** and **dew point**.

Relative humidity relates the actual water vapor present in the air to that which could be present in the air. Temperature largely determines the maximum amount of water vapor air can hold. Warm air can hold more water vapor than can cold air. *See* Figure 2-3.

Air with 100% relative humidity is said to be saturated, and less than 100% is unsaturated.

Dew point is the temperature to which air must be cooled to become saturated by the water already present in the air. *See* Figure 2-4.

Moisture can be added to the air by either evaporation or sublimation. Moisture is removed from the air by either condensation or sublimation. **Sublimation** is the process in which a solid material changes directly into a vapor, or a vapor changes into a solid without first becoming a liquid.

Continued

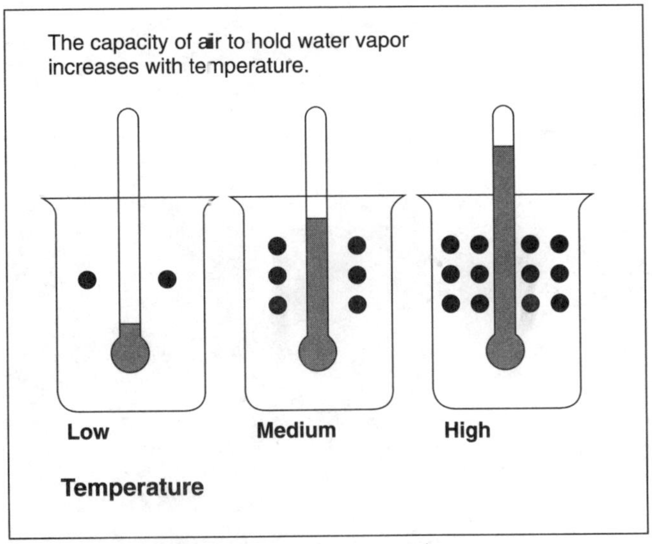

Figure 2-3. Capacity of air to hold water

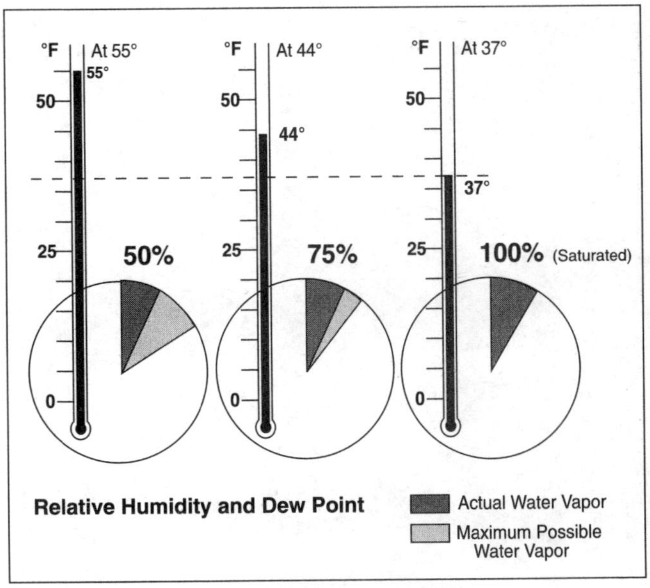

Figure 2-4. Relative humidity and dew point

Answers
2170 [C] 2312 [C]

When water vapor condenses on large objects such as leaves, windshields, or airplanes, it will form **dew**; and when it condenses on microscopic particles such as salt, dust, or combustion by-products (condensation nuclei), it will form clouds or fog.

To summarize, relative humidity can be increased either by lowering the air temperature or by increasing the amount of moisture in the air. If the temperature and dew point spread is small and decreasing, condensation is about to occur. If the temperature is above freezing, the weather most likely to develop will be fog or low clouds.

ALL

2207. What is meant by the term "dew point"?

A—The temperature at which air must be cooled to become saturated.
B—The temperature at which condensation and evaporation are equal.
C—The temperature which causes precipitation.

Dew point is the temperature to which air must be cooled to become saturated by the water vapor already present in the air. (PLT512) — AC 00-6A, Chapter 5

Answer (B) is incorrect because evaporation is not directly related to the dew point. Answer (C) is incorrect because dew or precipitation will form only when an object cools below the dew point of the surrounding air.

ALL

2169. Clouds, fog, or dew will always form when

A—water vapor condenses.
B—water vapor is present.
C—relative humidity reaches 100 percent.

As water vapor condenses or sublimates on condensation nuclei, liquid or ice particles begin to grow. Some condensation nuclei have an affinity for water and can induce condensation or sublimation even when air is almost, but not completely, saturated. (PLT512) — AC 00-6A, Chapter 5

Answer (B) is incorrect because the presence of water vapor does not result in clouds, fog, or dew unless condensation occurs. Answer (C) is incorrect because it is possible to have 100% humidity without the occurrence of condensation, which is necessary for clouds, fog, or dew to form.

Answers
2207 [A] 2169 [A]

Stability of the Atmosphere

Atmospheric stability is defined as the resistance of the atmosphere to vertical motion. A stable atmosphere resists any upward or downward movement. An unstable atmosphere allows an upward or downward disturbance to grow into a vertical (convective) current.

Atmospheric stability is determined by looking at how fast the air cools with increasing altitude. The standard lapse rate (cooling of the atmosphere) is 2°C per 1,000 feet elevation. This standard lapse rate would create moderate turbulence on a typical day. A stable day would be 1°C decrease for 1,000 feet altitude gain—smooth air with little vertical movement. An unstable day is when the decrease in temperature is 3°C per 1,000 feet, creating significant vertical air movement and unstable conditions which result in medium to heavy air turbulence. A 4°C per 1,000 feet decrease usually results in thunderstorms and wind gusts, the end result of an unstable day.

A stable layer of air would be associated with a temperature inversion (cold air below warm air). Warming from below, on the other hand, would decrease the stability of an air mass.

The conditions characteristic of stable air are: stratiform clouds and fog, continuous precipitation, smooth air, and fair to poor visibility in haze and smoke. The conditions characteristic of unstable air are: cumuliform clouds, showery precipitation, rough air (turbulence) and good visibility except in blowing obstructions. *See* Figure 2-5.

Unstable Air	Stable Air
Cumuliform clouds	Stratiform clouds and fog
Showery precipitation	Continuous precipitation
Rough air (turbulence)	Smooth air
Good visibility except in blowing obstructions	Fair to poor visibility in haze and smoke

Figure 2-5

ALL

2164. If an unstable air mass is forced upward, what type clouds can be expected?

A—Stratus clouds with little vertical development.
B—Stratus clouds with considerable associated turbulence.
C—Clouds with considerable vertical development and associated turbulence.

When unstable air is forced upward, the disturbance grows. Any resulting cloudiness shows extensive vertical development. (PLT511) — AC 00-6A, Chapter 6

ALL

2165. What are characteristics of a moist, unstable air mass?

A—Cumuliform clouds and showery precipitation.
B—Poor visibility and smooth air.
C—Stratiform clouds and showery precipitation.

Characteristics of a moist, unstable air mass include cumuliform clouds, showery precipitation, rough air (turbulence), and good visibility (except in blowing obstructions). (PLT511) — AC 00-6A, Chapter 6

ALL

2167. Which is a characteristic of stable air?

A—Cumuliform clouds.
B—Excellent visibility.
C—Restricted visibility.

Characteristics of a stable air mass include stratiform clouds and fog, continuous precipitation, smooth air, and fair to poor visibility in haze and smoke. (PLT173) — AC 00-6A, Chapter 6

Answers (A) and (B) are incorrect because these are characteristics of unstable air.

Answers

2164 [C] 2165 [A] 2167 [C]

Clouds

Stability determines which of two types of clouds will be formed: **cumuliform** or **stratiform**.

Cumuliform clouds are the billowy-type clouds having considerable vertical development, which enhances the growth rate of precipitation. They are formed in unstable conditions, and they produce showery precipitation made up of large water droplets. *See* Figure 2-6.

Stratiform clouds are the flat, more evenly based clouds formed in stable conditions. They produce steady, continuous light rain and drizzle made up of much smaller raindrops. *See* Figure 2-7.

Steady precipitation (in contrast to showery) preceding a front is an indication of stratiform clouds with little or no turbulence.

Clouds are divided into four families according to their height range: low, middle, high, and clouds with extensive vertical development. *See* Figure 2-8.

The first three families—low, middle, and high—are further classified according to the way they are formed. Clouds formed by vertical currents (unstable) are cumulus (heap) and are billowy in appearance. Clouds formed by the cooling of a stable layer are stratus (layered) and are flat and sheet-like in appearance. A further classification is the prefix "nimbo-" or suffix "-nimbus," which means raincloud.

High clouds, called **cirrus**, are composed mainly of ice crystals; therefore, they are least likely to contribute to structural icing (since it requires water droplets).

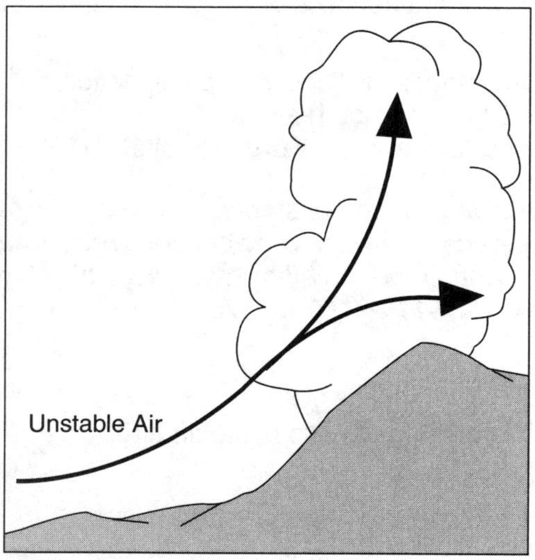

Figure 2-6. Cumulus clouds

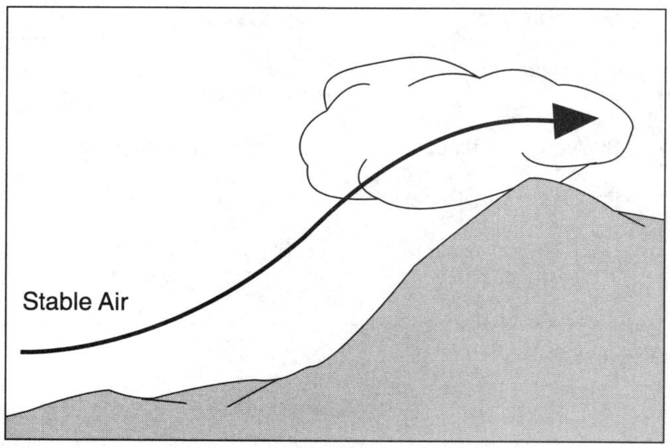

Figure 2-7. Stratiform clouds

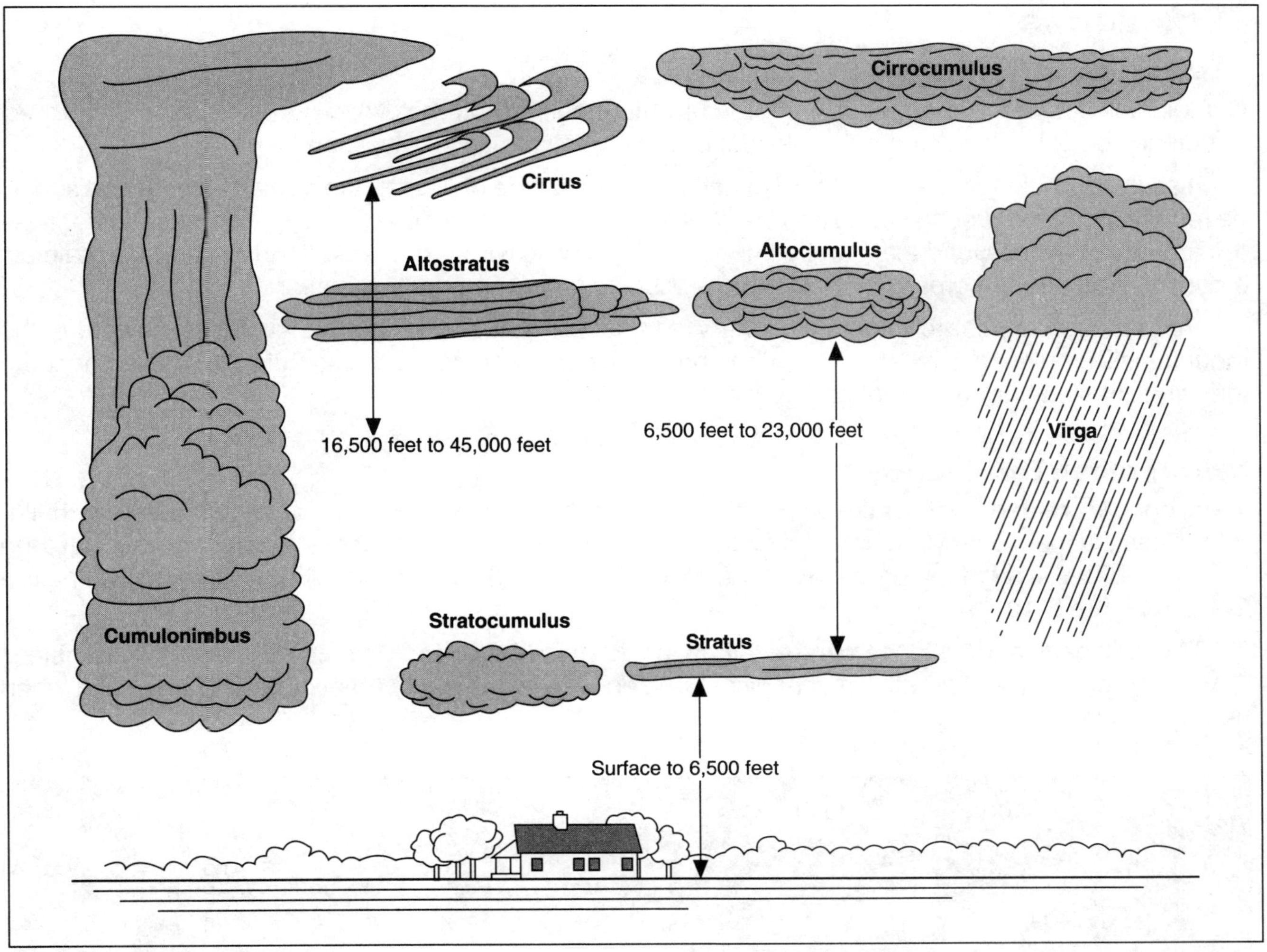

Figure 2-8. Cloud families

ALL

2166. The suffix 'nimbus,' used in naming clouds, means

A—a cloud with extensive vertical development.
B—a rain cloud.
C—a middle cloud containing ice pellets.

The prefix "nimbo-" or suffix "-nimbus" means rain cloud. (PLT192) — AC 00-6A, Chapter 7

Answer (A) is incorrect because these are cumulus. Answer (C) is incorrect because these are altostratus.

Answers
2166 [B]

Turbulence

Atmospheric Turbulence

Cumulus clouds are formed by convective currents (heating from below); therefore, a pilot can expect turbulence below or inside cumulus clouds, especially towering cumulus clouds.

If severe turbulence is encountered either inside or outside of clouds, the airplane's airspeed should be reduced to maneuvering speed and the pilot should attempt to maintain a level flight attitude because the amount of excess load that can be imposed on the wing will be decreased. Any attempt to maintain a constant altitude will greatly increase the stresses that are applied to the aircraft.

Rotor clouds appear to remain stationary, parallel the range, and stand a few miles leeward of the mountains. Turbulence is most frequent and most severe in and below the standing rotors just beneath the wave crests at or below mountaintop levels.

Mechanical Turbulence

Turbulence is created when air flows past any object on the ground, which can be large mountains, buildings, trees, bushes, or any other object on the ground. The stronger the wind is blowing, the stronger the turbulence. This is a hidden danger to any aircraft, especially those flying at slower speeds. *See* Figure 2-9 and Figure 2-10.

This mechanical turbulence can be felt as far downwind as 7 to 10 times the height of the object. Always expect and evaluate the mechanical turbulence in wind for any object and stay clear of it when near the ground.

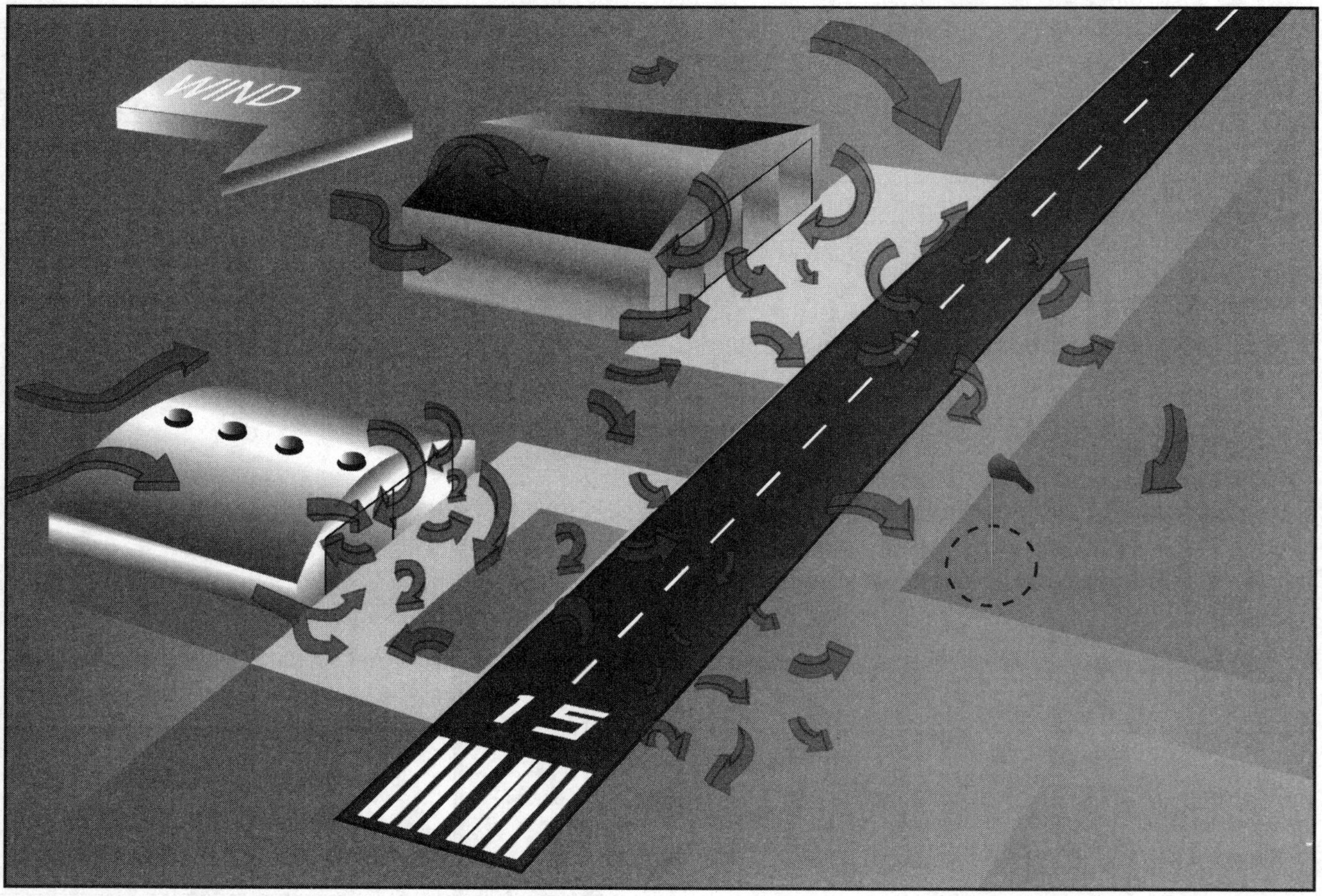

Figure 2-9. Turbulence caused by manmade obstructions

Figure 2-10. Turbulence encountered over and around mountainous regions

ALL
2159. One of the most dangerous features of mountain waves is the turbulent areas in and

A—below rotor clouds.
B—above rotor clouds.
C—below lenticular clouds.

Rotor clouds appear to remain stationary, parallel the range, and stand a few miles leeward of the mountains. Turbulence is most frequent and most severe in and below the standing rotors just beneath the wave crests at or below mountaintop levels. (PLT501) — AC 00-6A, Chapter 9

ALL
2160. What cloud types would indicate convective turbulence?

A—Cirrus clouds.
B—Nimbostratus clouds.
C—Towering cumulus clouds.

Towering cumulus signifies a relatively deep layer of unstable air. They show considerable vertical development and have billowing cauliflower tops. Showers can result from these clouds. Expect very strong turbulence, and perhaps some clear icing above the freezing level. (PLT192) — AC 00-6A, Chapter 7

Answers
2159 [A] 2160 [C]

Thunderstorms

Thunderstorms present many hazards to flying. Three conditions necessary to the formation of a thunderstorm are:

1. Sufficient water vapor;
2. An unstable lapse rate; and
3. An initial upward boost (lifting).

The initial upward boost can be caused by heating from below, frontal lifting, or by mechanical lifting (wind blowing air upslope on a mountain).

There are three stages of a thunderstorm: the cumulus, mature, and dissipating stages. *See* Figure 2-11.

The **cumulus stage** is characterized by continuous updrafts, and these updrafts create low-pressure areas. Thunderstorms reach their greatest intensity during the **mature stage**, which is characterized by updrafts and downdrafts inside the cloud. Precipitation inside the cloud aids in the development of these downdrafts, and the start of rain from the base of the cloud signals the beginning of the mature stage. The precipitation that evaporates before it reaches the ground is called **virga**. The **dissipating stage** of a thunderstorm is characterized predominantly by downdrafts.

Lightning is always associated with a thunderstorm.

Hail is formed inside thunderstorms by the constant freezing, melting, and refreezing of water as it is carried about by the up- and downdrafts.

A pilot should always expect the hazardous and invisible atmospheric phenomena called **wind shear turbulence** when operating anywhere near a thunderstorm (within 20 NM).

Thunderstorms that generally produce the most intense hazard to aircraft are called **squall-line thunderstorms**. These non-frontal, narrow bands of thunderstorms often develop ahead of a cold front.

Embedded thunderstorms are those that are obscured by massive cloud layers and cannot be seen.

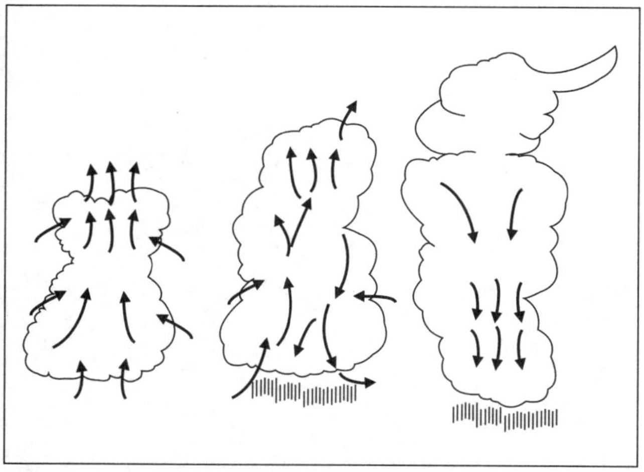

Figure 2-11. States of thunderstorms

ALL

2158. Thunderstorms which generally produce the most intense hazard to aircraft are

A—squall line thunderstorms.
B—air mass thunderstorms.
C—warm front thunderstorms.

A squall line is a non-frontal, narrow band of active thunderstorms. The line may be too long to easily detour and too wide and severe to penetrate. It often contains severe steady-state thunderstorms and presents the single, most intense weather hazard to aircraft. (PLT495) — AC 00-6A, Chapter 11

ALL

2172. What conditions are necessary for the formation of thunderstorms?

A—High humidity, lifting force, and unstable conditions.
B—High humidity, high temperature, and cumulus clouds.
C—Lifting force, moist air, and extensive cloud cover.

For a cumulonimbus cloud or thunderstorm to form, the air must have:

1. *Sufficient water vapor,*
2. *An unstable lapse rate, and*
3. *An initial upward boost (lifting) to start the storm process in motion.*

(PLT495) — AC 00-6A, Chapter 11

ALL

2173. Thunderstorms reach their greatest intensity during the

A—mature stage.
B—downdraft stage.
C—cumulus stage.

All thunderstorm hazards reach their greatest intensity during the mature stage. (PLT495) — AC 00-6A, Chapter 11

Answer (B) is incorrect because this is not one of the stages of a thunderstorm. Answer (C) is incorrect because the cumulus stage marks the beginning of the thunderstorm.

GLI

2311. Which is considered to be the most hazardous condition when soaring in the vicinity of thunderstorms?

A—Static electricity.
B—Lightning.
C—Wind shear and turbulence.

During the mature stage of a thunderstorm, updrafts and downdrafts in close proximity create strong vertical shears and a very turbulent environment. A lightning strike can puncture the skin of an aircraft and damage communication and navigation equipment. (PLT120) — AC 00-6A, Chapter 11

Answers

2158 [A]	2172 [A]	2173 [A]	2311 [C]

Wind Shear

Wind shear is defined as a change in wind direction and/or speed over a very short distance in the atmosphere. This can occur at any level of the atmosphere and can be detected by the pilot as a sudden change in airspeed.

Low-level (low-altitude) **wind shear** can be expected during strong temperature inversions, on all sides of a thunderstorm and directly below the cell. A pilot can expect a wind shear zone in a temperature inversion whenever the wind speed at 2,000 feet to 4,000 feet above the surface is at least 25 knots.

Low-level wind shear can also be found near frontal activity because winds can be significantly different in the two air masses which meet to form the front.

In warm front conditions, the most critical period is before the front passes. Warm front shear may exist below 5,000 feet for about 6 hours before surface passage of the front. The wind shear associated with a warm front is usually more extreme than that found in cold fronts.

The shear associated with cold fronts is usually found behind the front. If the front is moving at 30 knots or more, the shear zone will be 5,000 feet above the surface 3 hours after frontal passage.

Basically, there are two potentially hazardous shear situations—the loss of a tailwind or the loss of a headwind.

A tailwind may shear to either a calm or headwind component. The airspeed initially increases, the aircraft pitches up, and altitude increases. Lower than normal power would be required initially, followed by a further decrease as the shear is encountered, and then an increase as the glide slope is regained. A headwind may shear to a calm or tailwind component. Initially, the airspeed decreases, the aircraft pitches down, and altitude decreases. *See* Figure 2-12.

Some airports can report boundary winds as well as the wind at the tower. When a tower reports a boundary wind which is significantly different from the airport wind, there is a possibility of hazardous wind shear. When taking off or landing in calm wind under clear skies within a few hours before or after sunrise, be prepared for a temperature inversion near the ground. You can be relatively certain of a shear zone in the inversion if you know the wind at 2,000 to 4,000 feet is 25 knots or more. Allow a margin of airspeed above normal climb or approach speed to alleviate danger of stall in event of turbulence or sudden change in wind velocity. *See* Figure 2-13.

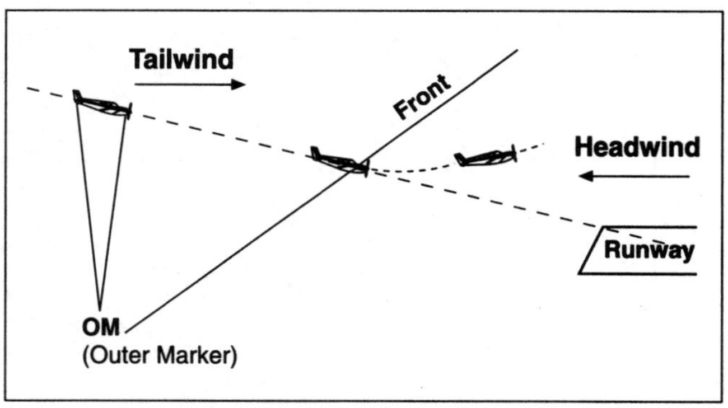

Figure 2-12. Tailwind shearing to headwind or calm

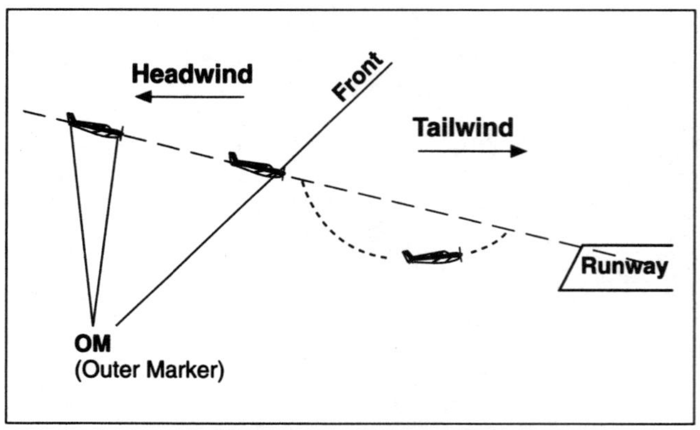

Figure 2-13. Headwind shearing to tailwind or calm

ALL

2162. Where does wind shear occur?

A—Only at higher altitudes.
B—Only at lower altitudes.
C—At all altitudes, in all directions.

Wind shear may be associated with either a wind shift or a wind speed gradient at any level in the atmosphere. (PLT518) — AC 00-6A, Chapter 9

ALL

2163. If a temperature inversion is encountered immediately after takeoff or during an approach to a landing, a potential hazard exists due to

A—wind shear.
B—strong surface winds.
C—strong convective currents.

When taking off or landing in calm wind under clear skies within a few hours before or after sunrise, be prepared for a temperature inversion near the ground. You can be relatively certain of a shear zone in the inversion if you know the wind at 2,000 to 4,000 feet is 25 knots or more. Allow a margin of airspeed above normal climb or approach speed to alleviate danger of stall in event of turbulence or sudden change in wind velocity. (PLT518) — AC 00-6A, Chapter 9

Answers

2162 [C] 2163 [A]

Weather Charts and Reports

Aviation Area Forecast (FA)

The aviation Area Forecast is a forecast of general weather conditions over an area the size of several states. It is used to determine forecast en route weather and to interpolate conditions at airports which do not have TAFs issued.

The Area Forecast is issued 3 times a day, and is comprised of four sections: a communications and product header section, a precautionary statement section, and two weather sections; a SYNOPSIS section and a VFR CLOUDS/WX section.

An example of an Area Forecast can be found in Figure 6 of the FAA Computer Testing Supplement.

Aviation Routine Weather Report (METAR)

An aviation routine weather report, or METAR, is an observation of current surface weather reported in a standard international format. An international weather reporting code is used for weather reports (METAR) and forecasts (TAFs) worldwide.

For aviation purposes, the ceiling is the lowest broken or overcast layer, or vertical visibility into an obscuration.

Key to Aerodrome Forecast (TAF)
and Aviation Routine Weather Report (METAR)

TAF KPIT 091730Z 091818 15005KT 5SM HZ FEW020 WS010/31022KT
FM 1930 30015G25KT 3SM SHRA OVC015 TEMPO 2022 1/2SM + TSRA OVC008CB
FM0100 27008KT 5SM SHRA BKN020 OVC040 PROB40 0407 1SM -RA BR
FM1015 18005KT 6SM -SHRA OVC020 BECMG 1315 P6SM NSW SKC

METAR KPIT 091955Z COR 22015G25KT 3/4SM R28L/2600FT TSRA OVC010CB 18/16 A2992 RMK SLP045 T01820159

Forecast	Explanation	Report
TAF	Message type: TAF-routine or TAF AMD-amended forecast, METAR-hourly, SPECI-special or TESTM-non-commissioned ASOS report	METAR
KPIT	ICAO location indicator	KPIT
091730Z	Issuance time: ALL times in UTC "Z", 2-digit date, 4-digit time	091955Z
091818	Valid period: 2-digit date, 2-digit beginning, 2-digit ending times	
	In U.S. **METAR**: CORrected ob; or AUTOmated ob for automated report with no human intervention; omitted when observer logs on	COR
15005KT	Wind: 3 digit true-north direction, nearest 10 degrees (or VaRiaBle); next 2-3 digits for speed and unit, KT (KMH or MPS); as needed, Gust and maximum speed; 00000KT for calm; for **METAR**, if direction varies 60 degrees or more, Variability appended, e.g. 180V260	22015G25KT
5SM	Prevailing visibility: in U.S., Statute Miles & fractions; above 6 miles in **TAF** Plus6SM. (Or, 4-digit minimum visibility in meters and as required, lowest value with direction)	3/4SM
	Runway Visual Range: R; 2-digit runway designator Left, Center, or Right as needed; "/"; Minus or Plus in U.S., 4-digit value, FeeT in U.S., (usually meters elsewhere); 4-digit value Variability 4-digit value (and tendency Down, Up or No change)	R28L/2600FT
HZ	Significant present, forecast and recent weather: see table (on back)	TSRA
FEW020	Cloud amount, height and type: SKy Clear 0/8, FEW >0/8-2/8, SCaTtered 3/8-4/8, BroKeN 5/8-7/8, OVerCast 8/8; 3-digit height in hundreds of ft; Towering CUmulus or CumulonimBus in **METAR**; in **TAF**, only CB. Vertical Visibility for obscured sky and height "VV004". More than 1 layer may be reported or forecast. In automated **METAR** reports only, CLeaR for "clear below 12,000 feet"	OVC010CB
	Temperature: degrees Celsius; first 2 digits, temperature "/" last 2 digits, dew-point temperature; Minus for below zero, e.g., M06	18/16
	Altimeter setting: indicator and 4 digits; in U.S., A-inches and hundredths; (Q-hectoPascals, e.g., Q1013)	A2992

Continued

Key to Aerodrome Forecast (TAF)
and Aviation Routine Weather Report (METAR) *Continued*

Forecast	Explanation	Report
WS010/31022KT	In U.S. **TAF**, non-convective low-level (≤ 2,000 ft) Wind Shear; 3-digit height (hundreds of ft); "/"; 3-digit wind direction and 2-3 digit wind speed above the indicated height, and unit, KT	
	In **METAR**, ReMarK indicator & remarks. For example: Sea-Level Pressure in hectoPascals & tenths, as shown: 1004.5 hPa; Temp/dew-point in tenths °C, as shown: temp. 18.2°C, dew-point 15.9°C	RMK SLP045 T01820159
FM1930	FroM and 2-digit hour and 2-digit minute **beginning** time: indicates significant change. Each FM starts on new line, indented 5 spaces.	
TEMPO 2022	TEMPOrary: changes expected for < 1 hour and in total, < half of 2-digit hour **beginning** and 2-digit hour **ending** time period	
PROB40 0407	PROBability and 2-digit percent (30 or 40): probable condition during 2-digit hour **beginning** and 2-digit hour **ending** time period	
BECMG 1315	BECoMinG: change expected during 2-digit hour **beginning** and 2-digit hour **ending** time period	

Table of Significant Present, Forecast and Recent Weather – Grouped in categories and used in the order listed below; or as needed in TAF, No Significant Weather.

QUALIFIER

Intensity or Proximity

'-' Light "no sign" Moderate '+' Heavy

VC Vicinity: but not at aerodrome; in U.S. **METAR**, between 5 and 10SM of the point(s) of observation; in U.S. **TAF**, 5 to 10SM from center of runway complex (elsewhere within 8000m)

Descriptor

MI Shallow	BC Patches	PR Partial	TS Thunderstorm
BL Blowing	SH Showers	DR Drifting	FZ Freezing

WEATHER PHENOMENA

Precipitation

DZ Drizzle	RA Rain	SN Snow	SG Snow grains
IC Ice crystals	PL Ice pellets	GR Hail	GS Small hail/snow pellets
UP Unknown precipitation in automated observations			

Obscuration

BR Mist (≥ 5/8SM)	FG Fog (< 5/8SM)	FU Smoke	VA Volcanic ash
SA Sand	HZ Haze	PY Spray	DU Widespread dust

Other

SQ Squall	SS Sandstorm	DS Duststorm	PO Well developed dust/sand whirls
FC Funnel cloud	+FC Tornado/waterspout		

– Explanation in parentheses "( )" indicate different worldwide practices.
– Ceiling is not specified; defined as the lowest broken or overcast layer, or the vertical visibility.
– NWS **TAFs** exclude turbulence, icing & temperature forecasts; NWS **METARs** exclude trend forecasts

January 1999
Aviation Weather Directorate

Department of Transportation
FEDERAL AVIATION ADMINISTRATION

Figure 2-14. TAF/METAR weather card

Terminal Aerodrome Forecast (TAF)

A Terminal Aerodrome Forecast (TAF) is a concise statement of the expected meteorological conditions at an airport during a specified period (usually 24 hours). TAFs use the same code used in the METAR weather reports.

TAFs are issued in the following format:

TYPE / LOCATION / ISSUANCE TIME / VALID TIME / FORECAST

Note: The "/" above are for separation purposes and do not appear in the actual TAFs.

Radar Weather Reports

Radar Weather Reports (SDs) are of special interest to the pilot because they indicate the location of precipitation with type and intensity.

The Radar Summary Chart will yield a three-dimensional view of clouds and precipitation when used in conjunction with other charts and reports. Since radar detects only drops or ice particles of precipitation size, it does not detect clouds or fog. It is the only chart that shows lines and cells of hazardous thunderstorms.

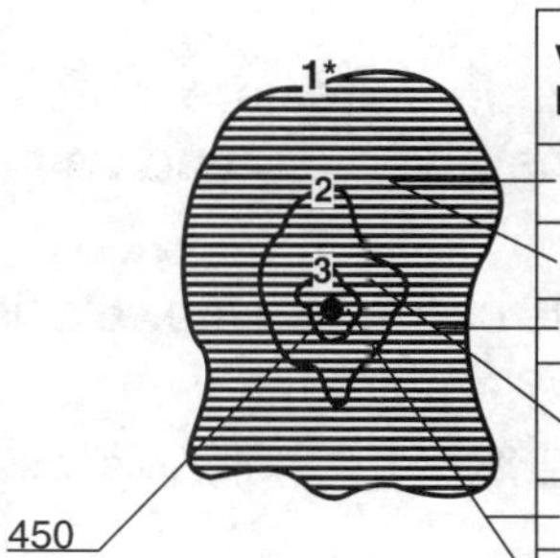

Highest precipitation top in area in hundreds of feet MSL (45,000 Feet MSL)

VIP Level	Echo Intensity	Precipitation Intensity	Rainfall Rate in/hr Stratiform	Rainfall Rate in/hr Convective
1	Weak	Light	Less than 0.1	Less than 0.2
2	Moderate	Moderate	0.1 – 0.5	0.2 – 1.1
3	Strong	Heavy	0.5 – 1.0	1.1 – 2.2
4	Very strong	Very heavy	1.0 – 2.0	2.2 – 4.5
5	Intense	Intense	2.0 – 5.0	4.5 – 7.1
6	Extreme	Extreme	More than 5.0	More than 7.1

*The numbers representing the intensity level do not appear on the chart. Beginning from the first contour line, bordering the area, the intensity level is 1–2, second contour is 3–4, and the third contour is 5–6.

Symbols Used on Charts

Symbol	*Meaning*
R	Rain
RW	Rain shower
Hail	Hail
S	Snow
PL	Ice pellets
SW	Snow shower
L	Drizzle
T	Thunderstorm
ZR, ZL	Freezing precipitation
NE	No echoes observed
NA	Observations unavailbale
OM	Out for Maintenance
STC	STC ON – all precipitation may not be seen
ROBEPS	Radar Operating Below Performance Standards
RHINO	Range Height Indicator Not Operating

Symbol	*Meaning*
+	Intensity increasing or new echo
–	Intensity decreasing
No Symbol	No change in intensity
35	Cell movement to NE at 35 knots
	Line or area movement to East at 20 knots
LM	Little movement
MA	Echoes mostly aloft
PA	Echoes partly aloft

Symbol	*Meaning*
	Line of echoes
SLD	8/10 or greater coverage in a line
WS999	Severe thunderstorm watch
WT999	Tornado watch
LEWP	Line echo wave pattern
HOOK	Hook echo
BWER	Bounded weak echo region
PCLL	Persistent cell
FNLN	Fine line

Figure 2-15

Areas from which precipitation echoes were received are shown on the Radar Summary Chart, with echo intensity indicated by contours as shown in Figure 2-15. Echo heights are displayed in hundreds of feet MSL. Tops are entered above a short line, while any available bases are entered below a short line. Movement of an area of echoes is indicated by a shaft and barb combination with the shaft indicating the direction, and the barbs indicating the speed. A whole barb is 10 knots, a half barb is 5 knots, and a pennant is 50 knots. Individual cell movement is indicated by an arrow that shows the direction of movement with the speed in knots entered as a number. Severe weather watch areas are outlined by heavy dashed lines, usually in the form of a large rectangular box.

Winds and Temperatures Aloft Forecast

The winds and temperatures aloft forecast (FD) shows wind direction, wind velocity, and the temperature that is forecast to exist at specified levels. This is used for flight planning, determining stability, and understanding how hard the wind is blowing above you. It is a most useful tool for sport pilots.

If you are flying in the early morning and the wind is calm at the airport, and you determined the winds are blowing 30 MPH 3,000 feet above you, you are in an inversion and there is a good probability of wind shear above you. You also could expect the higher winds to reach the ground sometime in the morning or possibly during your flight. The winds and temperatures aloft forecast is necessary for determining if you will have a headwind or a tailwind for your cross-country flight.

No winds are forecast for a level within 1,500 feet of the surface, and no temperatures are forecast for the 3,000-foot level or for a level within 2,500 feet of the surface. This is where you depend on the terminal area forecast (TAF) and airport observations (METAR).

The upper air wind direction is shown in tens of degrees with reference to true north, and the velocity is shown in knots. Temperatures are in degrees Celsius, and are negative above 24,000 feet.

Each 6-digit group includes wind and temperature. For example, the entry "2032+15" indicates wind from 200° true north at 32 knots, temperature 15°C.

A forecast wind from 160 degrees true at 25 knots with a temperature at minus 2°C would be encoded "1625-02"; "9900" indicates wind light and variable, speed less than 5 knots.

FD WBC 151745
DATA BASED ON 151200Z
VALID 1600Z FOR USE 1800–0300Z. TEMPS NEG ABV 24000

FT	3000	6000	9000	12000	18000	24000	30000	34000	39000
ALS			2420	2635-08	2535-18	2444-30	245945	246755	246862
AMA		2714	2725+00	2625-04	2531-15	2542-27	265842	256352	256762
DEN			2321-04	2532-08	2434-19	2441-31	235347	236056	236262
HLC		1707-01	2113-03	2219-07	2330-17	2435-30	244145	244854	245561
MKC	0507	2006+03	2215-01	2322-06	2338-17	2348-29	236143	237252	238160
STL	2113	2325+07	2332+02	2339-04	2356-16	2373-27	239440	730649	731960

The wind in the Denver area at 9,000 feet is 230 degrees true, 21 knots, and the temperature is minus 4°C. The temperature drops 4 degrees from 9,000 to 12,000 feet telling us it is a stable day. This would be for the time of day 1800 – 0300Z or (1800 minus 7 hours) 11 a.m. to (0300 minus 7 hours) 8 p.m.

ALL

2155. For aviation purposes, ceiling is defined as the height above the Earth's surface of the

A—lowest reported obscuration and the highest layer of clouds reported as overcast.
B—lowest broken or overcast layer or vertical visibility into an obscuration.
C—lowest layer of clouds reported as scattered, broken, or thin.

For aviation purposes, the ceiling is defined as the height (AGL) of the lowest broken or overcast layer aloft or vertical visibility into an obscuration. (PLT026) — AC 00-45

ALL

2156. (Refer to Figure 53.) The wind direction and velocity at KJFK is from

A—180°; true at 4 knots.
B—180°; magnetic at 4 knots.
C—040°; true at 18 knots.

The wind is reported as a five-digit group (six digits if speed is over 99 knots). The first three digits is the direction the wind is blowing from in tens of degrees relative to true north, or "VRB" if the direction is variable. The next two digits is the speed in knots, or if over 99 knots, the next three digits. If the wind is gusty, it is reported as a "G" after the speed followed by the highest gust reported. The abbreviation "KT" is appended to denote the use of knots for wind speed. The wind group for KJFK is 18004KT which means the wind is from 180° at 4 knots. (PLT059) — AC 00-45

ALL

2281. From which primary source should information be obtained regarding expected weather at the estimated time of arrival if your destination has no Terminal Forecast?

A—Low-Level Prognostic Chart.
B—Weather Depiction Chart.
C—Area Forecast.

An Area Forecast (FA) is used to determine forecast enroute weather and to interpolate conditions at airports which do not have Terminal Forecasts (FT) issued. (PLT291) — AC 00-45

ALL

2209. (Refer to Figure 53.) Which of the reporting stations have VFR weather?

A—All.
B—KINK, KBOI, and KJFK.
C—KINK, KBOI, and KLAX.

VFR conditions are a ceiling 1,000 feet or greater and/or visibility greater than 3 miles.

KINK: Visibility is 15 statute miles and the sky is clear: VFR.

KBOI: Visibility is 30 statute miles and sky is scattered at 15,000 feet: VFR.

KLAX: Visibility is 6 statute miles with mist and the sky is scattered at 700 feet (SCT does not constitute a ceiling): VFR.

KMDW: Visibility is 1-1/2 statute miles with rain and the ceiling is overcast at 700 feet: IFR.

KJFK: Visibility is 1/2 statute mile with fog and the ceiling is overcast at 500 feet: IFR.

(PLT059) — AC 00-45

ALL

2191. What information is provided by the Radar Summary Chart that is not shown on other weather charts?

A—Lines and cells of hazardous thunderstorms.
B—Ceilings and precipitation between reporting stations.
C—Areas of cloud cover and icing levels within the clouds.

Radar detects objects of precipitation size or greater. Based on the amount of precipitation radar detects, it can determine if it is a thunderstorm. (PLT353) — AC 00-45

Answers (B) and (C) are incorrect because it is not designed to detect ceilings and types of clouds.

Answers

2155 [B] 2156 [A] 2281 [C] 2209 [C] 2191 [A]

Inflight Weather Advisories

Inflight Weather Advisories advise enroute pilots of the possibility of encountering hazardous flying conditions that may not have been forecast at the time of the preflight weather briefing. They are available on NAVAIDs that have HIWAS capability, depicted on sectional charts with an "H" in the upper right corner of the identification box, or by contacting FSS or EFAS typically on 122.0.

AIRMETs (WA) contain information on weather that may be hazardous to single engine, other light aircraft, and VFR pilots. The items covered are moderate icing or turbulence, sustained winds of 30 knots or more at the surface, widespread areas of IFR conditions, and extensive mountain obscurement.

SIGMETs (WS) advise of weather potentially hazardous to all aircraft. The items covered are severe icing, severe or extreme turbulence, and widespread sandstorms, dust storms or volcanic ash lowering visibility to less than 3 miles.

SIGMETs and AIRMETs are broadcast upon receipt and at 30-minute intervals (H + 15 and H + 45) during the first hour. If the advisory is still in effect after the first hour, an alert notice will be broadcast. Pilots may contact the nearest FSS to ascertain whether the advisory is pertinent to their flights.

Convective SIGMETs (WST) cover weather developments such as tornadoes, lines of thunderstorms, and embedded thunderstorms, and they also imply severe or greater turbulence, severe icing, and low-level wind shear. When a SIGMET forecasts embedded thunderstorms, it indicates that the thunderstorms are obscured by massive cloud layers and cannot be seen. Convective SIGMET bulletins are issued hourly at H + 55. Unscheduled convective SIGMETs are broadcast upon receipt and at 15-minute intervals for the first hour (H + 15; H + 30; H + 45).

ALL

2157. SIGMET's are issued as a warning of weather conditions hazardous to which aircraft?

A—Small aircraft only.
B—Large aircraft only.
C—All aircraft.

A SIGMET advises of weather potentially hazardous to all aircraft. (PLT290) — AC 00-45

Answer (A) is incorrect because these are AIRMETs. Answer (B) is incorrect because there is no inflight weather advisory specific to large aircraft only.

Answers
2157 [C]

Obtaining a Telephone Weather Briefing

When telephoning a weather briefing facility for preflight weather information, pilots should:

- Identify themselves as sport pilots (so the briefer will know they will be flying below 10,000 feet MSL, daytime only, and slower than 120 knots);
- State that they will be flying VFR in light-sport aircraft and what the category is, if not an airplane;
- State the intended route, destination, and type of aircraft;
- Request a standard briefing to get a "complete" weather briefing;
- Request an abbreviated briefing to supplement mass disseminated data or when only one or two items are needed; and
- Request an outlook briefing whenever the proposed departure time is 6 or more hours from the time of briefing.

ALL

2153. What should pilots state initially when telephoning a weather briefing facility for preflight weather information?

A—Tell the number of occupants on board.
B—Identify themselves as pilots.
C—State their total flight time.

When requesting a briefing, pilots should identify themselves and provide as much information regarding the proposed flight as possible. The information received will depend on the type of briefing requested. The following would be helpful to the briefer.

1. *Type of flight, VFR*
2. *Aircraft number or pilot's name*
3. *Aircraft type*
4. *Departure point*
5. *Route of flight*
6. *Destination*
7. *Flight altitude(s)*
8. *Estimated time of departure*
9. *Estimated time en route or estimated time of arrival*

(PLT514) — FAA-H-8083-25

Answers (A) and (C) are incorrect because these are necessary when filing a flight plan, but not when obtaining a weather briefing.

ALL

2154. Which type weather briefing should a pilot request, when departing within the hour, if no preliminary weather information has been received?

A—Outlook briefing.
B—Abbreviated briefing.
C—Standard briefing.

You should request a standard briefing any time you are planning a flight and you have not received a previous briefing or preliminary information through mass dissemination media, e.g., TWEB, PATWAS, VRS, etc. (PLT514) — AC 00-45

Answer (A) is incorrect because an outlook briefing should be requested when a planned departure is 6 or more hours away. Answer (B) is incorrect because an abbreviated briefing should be requested when a departure has been delayed or when specific weather information is needed to update the previous briefing.

ALL

2210. To update a previous weather briefing, a pilot should request

A—an abbreviated briefing.
B—a standard briefing.
C—an outlook briefing.

Request an abbreviated briefing when you need information to supplement mass disseminated data, update a previous briefing, or when you need only one or two specific items. (PLT514) — AC 00-45

Answers

2153 [B] 2154 [C] 2210 [A]

Chapter 3
Flight Physiology

Fitness for Flight

Pilot performance can be seriously degraded by a number of physiological factors. While some of the factors may be beyond the control of the pilot, awareness of cause and effect can help minimize any adverse effects. The safest rule is not to fly while suffering from any illness. If this rule is considered too stringent for a particular illness, the pilot should contact an Aviation Medical Examiner for advice.

Pilot performance can be seriously degraded by both prescribed and over-the-counter medications, as well as by the medical conditions for which they are taken. Flying is almost always precluded while using prescription analgesics since these drugs may cause side effects such as mental confusion, dizziness, headaches, nausea, and vision problems. Depressants, including antihistamines, lower blood pressure, reduce mental processing, and slow motor and reaction responses.

Dehydration is the term given to a critical loss of water from the body. As a pilot, flying for long periods in hot summer temperatures or at high altitudes increases the susceptibility of dehydration since the dry air at altitude tends to increase the rate of water loss from the body. If this fluid is not replaced, fatigue progresses to dizziness, weakness, nausea, tingling of hands and feet, abdominal cramps, and extreme thirst.

Hypoxia, a state of oxygen deficiency, impairs functions of the brain and other organs. Headache, drowsiness, dizziness, and euphoria are all symptoms of hypoxia.

Hyperventilation, a deficiency of carbon dioxide within the body, can be the result of rapid or extra deep breathing due to emotional tension, anxiety, or fear. As hyperventilation "blows off" excessive carbon dioxide from the body, a pilot can experience symptoms of light-headedness, suffocation, drowsiness, tingling in the extremities, and coolness and react to them with even greater hyperventilation. Incapacitation can eventually result from uncoordination, disorientation, and painful muscle spasms. Finally, unconsciousness can occur. Symptoms will subside after the rate and depth of breathing are brought under control. A pilot should be able to overcome the symptoms or avoid future occurrences of hyperventilation by talking aloud, breathing into a bag, or slowing the breathing rate.

Carbon monoxide is a colorless, odorless, and tasteless gas contained in exhaust fumes. Symptoms of carbon monoxide poisoning include headache, drowsiness, or dizziness. Large accumulations of carbon monoxide in the human body result in a loss of muscular power. Susceptibility increases as altitude increases. A pilot who detects symptoms of carbon monoxide poisoning should immediately shut off the heater and open air vents.

Various complex motions, forces, and visual scenes encountered in flight may result in misleading information being sent to the brain by various sensory organs. Spatial disorientation may result if these body signals are used to interpret flight attitude.

ALL

2090. If advice is needed concerning possible flight with an illness, a pilot should contact

A—an Aviation Medical Examiner.
B—their family doctor.
C—the nearest hospital.

The safest rule is not to fly while suffering from any illness. If this rule is considered too stringent for a particular illness, the pilot should contact an Aviation Medical Examiner for advice. (PLT098) — AIM ¶8-1-1

Answers
2090 [A]

ALL

2202. Which will almost always affect your ability to fly?

A—Over-the-counter analgesics and antihistamines.
B—Antibiotics and anesthetic drugs.
C—Prescription analgesics and antihistamines.

Pilot performance can be seriously degraded by both prescribed and over-the-counter medications, as well as by the medical conditions for which they are taken. Flying is almost always precluded while using prescription analgesics since these drugs may cause side effects such as mental confusion, dizziness, headaches, nausea, and vision problems. Depressants, including antihistamines, lower blood pressure, reduce mental processing, and slow motor and reaction responses. (PLT098) — FAA-H-8083-25

Answer (A) is incorrect because over-the-counter analgesics such as aspirin, Tylenol and Advil, and antihistamines have few side effects when taken in the correct dosage. Answer (B) is incorrect because these drugs do not typically limit flying (but the medical conditions for which they are being taken may indeed limit flying activities).

ALL

2089. As a pilot, flying for long periods in hot summer temperatures increases the susceptibility of dehydration since the

A—dry air at altitude tends to increase the rate of water loss from the body.
B—moist air at altitude helps retain the body's moisture.
C—temperature decreases with altitude.

As a pilot, flying for long periods in hot summer temperatures or at high altitudes increases the susceptibility of dehydration since the dry air at altitude tends to increase the rate of water loss from the body. If this fluid is not replaced, fatigue progresses to dizziness, weakness, nausea, tingling of hands and feet, abdominal cramps, and extreme thirst. (PLT098) — FAA-H-8083-25

ALL

2091. A pilot should be able to overcome the symptoms or avoid future occurrences of hyperventilation by

A—closely monitoring the flight instruments to control the airplane.
B—slowing the breathing rate or breathing into a bag.
C—increasing the breathing rate in order to increase lung ventilation.

The symptoms of hyperventilation subside within a few minutes after the rate and depth of breathing are consciously brought back under control. The buildup of carbon dioxide in the body can be hastened by controlled breathing in and out of a paper bag held over the nose and mouth. Talking aloud often helps, while normally-paced breathing at all times prevents hyperventilation. (PLT332) — AIM ¶8-1-3

ALL

2092. As hyperventilation progresses a pilot can experience

A—decreased breathing rate and depth.
B—heightened awareness and feeling of well being.
C—symptoms of suffocation and drowsiness.

As hyperventilation "blows off" excessive carbon dioxide from the body, a pilot can experience symptoms of lightheadedness, suffocation, drowsiness, tingling in the extremities, and coolness and react to them with even greater hyperventilation. Incapacitation can eventually result from uncoordination, disorientation, and painful muscle spasms. Finally, unconsciousness can occur. (PLT332) — AIM ¶8-1-3

ALL

2093. To overcome the symptoms of hyperventilation, a pilot should

A—swallow or yawn.
B—slow the breathing rate.
C—increase the breathing rate.

The symptoms of hyperventilation subside within a few minutes after the rate and depth of breathing are consciously brought back under control. The buildup of carbon dioxide in the body can be hastened by controlled breathing in and out of a paper bag held over the nose and mouth. (PLT332) — AIM ¶8-1-3

Answers

2202 [C] 2089 [A] 2091 [B] 2092 [C] 2093 [B]

ALL

2094. Which would most likely result in hyperventilation?

A—Emotional tension, anxiety, or fear.
B—The excessive consumption of alcohol.
C—An extremely slow rate of breathing and insufficient oxygen.

Hyperventilation is most likely to occur during periods of stress or anxiety. (PLT332) — AIM ¶8-1-3

ALL

2095. Large accumulations of carbon monoxide in the human body result in

A—tightness across the forehead.
B—loss of muscular power.
C—an increased sense of well-being.

A large accumulation of carbon monoxide in the body results in loss of muscular power, vomiting, convulsions, and coma. (PLT097) — FAA-H-8083-25

Answers (A) and (C) are incorrect because hypoxia gives you an increased sense of well-being and stress could result in tightness across the forehead.

ALL

2097. A state of temporary confusion resulting from misleading information being sent to the brain by various sensory organs is defined as

A—spatial disorientation.
B—hyperventilation.
C—hypoxia.

Disorientation, or vertigo, is actually a state of temporary spatial confusion resulting from misleading information sent to the brain by various sensory organs. (PLT334) — AIM ¶8-1-5

Answers

2094 [A]	2095 [B]	2097 [A]

Visual Illusions

Rapid acceleration during takeoff can create the **illusion** of being in a nose-up attitude, and a disoriented pilot will push the aircraft into a nose-low, or dive attitude. Rapid deceleration caused by a quick reduction of the throttles can have the opposite effect, with the disoriented pilot pulling the aircraft into a nose-up, or stall attitude.

An upsloping runway, upsloping terrain, or both, can create the illusion that the aircraft is at a higher altitude than it actually is. The pilot who does not recognize this illusion will fly a lower approach. A downsloping runway, downsloping approach terrain, or both can have the opposite effect.

Rain on the windscreen can create the illusion of greater height, and atmospheric haze causes the illusion of being a greater distance from the runway. The pilot who does not recognize these illusions will fly a lower approach.

ALL

2098. What effect does haze have on the ability to see traffic or terrain features during flight?

A—Haze causes the eyes to focus at infinity.
B—The eyes tend to overwork in haze and do not detect relative movement easily.
C—All traffic or terrain features appear to be farther away than their actual distance.

Atmospheric haze can create the illusion of being at a greater distance from objects on the ground and in the air. (PLT280) — AIM ¶8-1-5

ALL

2240. Haze creates which of the following atmospheric illusions?

A—Being at a greater distance from the runway.
B—Being at a closer distance from the runway.
C—Haze creates no atmospheric illusions.

Atmospheric haze can create the illusion of being at a greater distance from objects on the ground and in the air. (PLT280) — AIM ¶8-1-5

Answers

2098 [C] 2240 [A]

Aeronautical Decision Making

The pilot is responsible for determining whether he/she is fit to fly for a particular flight. Most preventable accidents have one common factor: human error, rather than a mechanical malfunction. The poor judgment chain, sometimes referred to as the "error chain," is a term used to describe this concept of contributing factors in a human factors-related accident. Breaking one link in the chain normally is all that is necessary to change the outcome of the sequence of events. Good aeronautical decision making (ADM) is necessary to prevent human error. ADM is a systematic approach to the mental process used by aircraft pilots to consistently determine the best course of action in response to a given set of circumstances.

The ADM process addresses all aspects of decision making in the cockpit and identifies the steps involved in good decision making. Steps for good decision making are:

1. Identifying personal attitudes hazardous to safe flight.
2. Learning behavior modification techniques.
3. Learning how to recognize and cope with stress.
4. Developing risk assessment skills.
5. Using all resources in a multicrew situation.
6. Evaluating the effectiveness of one's ADM skills.

There are a number of classic behavioral traps into which pilots have been known to fall. Pilots, particularly those with considerable experience, as a rule always try to complete a flight as planned, please passengers, meet schedules, and generally demonstrate that they have the "right stuff." These tendencies ultimately may lead to practices that are dangerous and often illegal, and may lead to a mishap. All experienced pilots have fallen prey to, or have been tempted by, one or more of these tendencies in their flying careers. These dangerous tendencies or behavior patterns, which must be identified and eliminated, include:

1. **Peer Pressure.** Poor decision making based upon emotional response to peers rather than evaluating a situation objectively.
2. **Mind Set.** The inability to recognize and cope with changes in the situation different from those anticipated or planned.
3. **Get-There-Itis.** This tendency, common among pilots, clouds the vision and impairs judgment by causing a fixation on the original goal or destination combined with a total disregard for any alternative course of action.
4. **Duck-Under Syndrome.** The tendency to sneak a peek by descending below minimums during an approach. Based on a belief that there is always a built-in "fudge" factor that can be used or on an unwillingness to admit defeat and shoot a missed approach.
5. **Scud Running.** Pushing the capabilities of the pilot and the aircraft to the limits by trying to maintain visual contact with the terrain while trying to avoid physical contact with it. This attitude is characterized by the old pilot's joke: "If it's too bad to go IFR, we'll go VFR."
6. **Continuing Visual Flight Rules (VFR)** into instrument conditions often leads to spatial disorientation or collision with ground/obstacles. It is even more dangerous if the pilot is not instrument qualified or current.
7. **Getting Behind the Aircraft.** Allowing events or the situation to control your actions rather than the other way around. Characterized by a constant state of surprise at what happens next.

Continued

8. **Loss of Positional or Situation Awareness.** Another case of getting behind the aircraft which results in not knowing where you are, an inability to recognize deteriorating circumstances, and/or the misjudgment of the rate of deterioration.
9. **Operating Without Adequate Fuel Reserves.** Ignoring minimum fuel reserve requirements, either VFR or Instrument Flight Rules (IFR), is generally the result of overconfidence, lack of flight planning, or ignoring the regulations.
10. **Descent Below the Minimum Enroute Altitude.** The duck-under syndrome (mentioned above) manifesting itself during the enroute portion of an IFR flight.
11. **Flying Outside the Envelope.** Unjustified reliance on the (usually mistaken) belief that the aircraft's high performance capability meets the demands imposed by the pilot's (usually overestimated) flying skills.
12. **Neglect of Flight Planning, Preflight Inspections, Checklists, etc.** Unjustified reliance on the pilot's short and long term memory, regular flying skills, repetitive and familiar routes, etc.

Each ADM student should take the Self-Assessment Hazardous Attitude Inventory Test in order to gain a realistic perspective on his/her attitudes toward flying. The inventory test requires the pilot to provide a response which most accurately reflects the reasoning behind his/her decision. The pilot must choose one of the five given reasons for making that decision, even though the pilot may not consider any of the five choices acceptable. The inventory test presents extreme cases of incorrect pilot decision making in an effort to introduce the five types of hazardous attitudes.

ADM addresses the following five hazardous attitudes:

1. **Antiauthority (don't tell me!).** This attitude is found in people who do not like anyone telling them what to do. In a sense they are saying "no one can tell me what to do." They may be resentful of having someone tell them what to do or may regard rules, regulations, and procedures as silly or unnecessary. However, it is always your prerogative to question authority if you feel it is in error. The antidote for this attitude is: Follow the rules. They are usually right.
2. **Impulsivity (do something quickly!)** is the attitude of people who frequently feel the need to do something—*anything*—immediately. They do not stop to think about what they are about to do, they do not select the best alternative, and they do the first thing that comes to mind. The antidote for this attitude is: Not so fast. Think first.
3. **Invulnerability (it won't happen to me).** Many people feel that accidents happen to others, but never to them. They know accidents can happen, and they know that anyone can be affected. They never really feel or believe that they will be personally involved. Pilots who think this way are more likely to take chances and increase risk. The antidote for this attitude is: It could happen to me.
4. **Macho (I can do it).** Pilots who are always trying to prove that they are better than anyone else are thinking "I can do it—I'll show them." Pilots with this type of attitude will try to prove themselves by taking risks in order to impress others. While this pattern is thought to be a male characteristic, women are equally susceptible. The antidote for this attitude is: taking chances is foolish.
5. **Resignation (what's the use?).** Pilots who think "what's the use?" do not see themselves as being able to make a great deal of difference in what happens to them. When things go well, the pilot is apt to think that's good luck. When things go badly, the pilot may feel that "someone is out to get me," or attribute it to bad luck. The pilot will leave the action to others, for better or worse. Sometimes, such pilots will even go along with unreasonable requests just to be a "nice guy." The antidote for this attitude is: I'm not helpless. I can make a difference.

Hazardous attitudes which contribute to poor pilot judgment can be effectively counteracted by redirecting that hazardous attitude so that appropriate action can be taken. Recognition of hazardous thoughts is the first step in neutralizing them in the ADM process. Pilots should become familiar with a means of counteracting hazardous attitudes with an appropriate antidote thought. When a pilot recognizes a thought as hazardous, the pilot should label that thought as hazardous, then correct that thought by stating the corresponding antidote.

If you hope to succeed at reducing stress associated with crisis management in the air or with your job, it is essential to begin by making a personal assessment of stress in all areas of your life. Good **cockpit stress management** begins with good life stress management. Many of the stress coping techniques practiced for life stress management are not usually practical in flight. Rather, you must condition yourself to relax and think rationally when stress appears. The following checklist outlines some thoughts on cockpit stress management.

1. Avoid situations that distract you from flying the aircraft.
2. Reduce your workload to reduce stress levels. This will create a proper environment in which to make good decisions.
3. If an emergency does occur, be calm. Think for a moment, weigh the alternatives, then act.
4. Maintain proficiency in your aircraft; proficiency builds confidence. Familiarize yourself thoroughly with your aircraft, its systems, and emergency procedures.
5. Know and respect your own personal limits.
6. Do not let little mistakes bother you until they build into a big thing. Wait until after you land, then "debrief" and analyze past actions.
7. If flying is adding to your stress, either stop flying or seek professional help to manage your stress within acceptable limits.

The **DECIDE Model**, comprised of a six-step process, is intended to provide the pilot with a logical way of approaching decision making. The six elements of the DECIDE Model represent a continuous loop decision process which can be used to assist a pilot in the decision making process when he/she is faced with a change in a situation that requires a judgment. This DECIDE Model is primarily focused on the intellectual component, but can have an impact on the motivational component of judgment as well. If a pilot practices the DECIDE Model in all decision making, its use can become very natural and could result in better decisions being made under all types of situations.

1. **D**etect. The decisionmaker detects the fact that change has occurred.
2. **E**stimate. The decisionmaker estimates the need to counter or react to the change.
3. **C**hoose. The decisionmaker chooses a desirable outcome (in terms of success) for the flight.
4. **I**dentify. The decisionmaker identifies actions which could successfully control the change.
5. **D**o. The decisionmaker takes the necessary action.
6. **E**valuate. The decisionmaker evaluates the effect(s) of his/her action countering the change.

ALL
2086. What is the one common factor which affects most preventable accidents?

A—Human error.
B—Mechanical malfunction.
C—Structural failure.

Most preventable accidents have one common factor: human error, rather than a mechanical malfunction. (PLT271) — AC 60-22

ALL
2073. A series of judgmental errors which can lead to a human factors-related accident is sometimes referred to as the

A—error chain.
B—course of action.
C—DECIDE model.

The poor judgment chain, sometimes referred to as the "error chain," is a term used to describe this concept of contributing factors in a human factors-related accident. Breaking one link in the chain normally is all that is necessary to change the outcome of the sequence of events. (PLT271) — FAA-H-8083-25

ALL
2085. Which of the following is the first step of the Decide Model for effective risk management and Aeronautical Decision Making?

A—Identify.
B—Detect.
C—Evaluate.

The DECIDE Model, comprised of a six-step process, is intended to provide the pilot with a logical way of approaching decision making: Detect, Estimate, Choose, Identify, Do, and Evaluate. (PLT022) — AC 60-22

ALL
2074. What are some of the hazardous attitudes dealt with in Aeronautical Decision Making (ADM)?

A—Risk management, stress management, and risk elements.
B—Poor decision making, situational awareness, and judgment.
C—Antiauthority (don't tell me), impulsivity (do something quickly without thinking), macho (I can do it).

ADM addresses the following five hazardous attitudes: Antiauthority (don't tell me!), Impulsivity (do something quickly!), Invulnerability (it won't happen to me), Macho (I can do it), Resignation (what's the use?). (PLT103) — AC 60-22

ALL
2075. What is the first step in neutralizing a hazardous attitude in the ADM process?

A—Dealing with improper judgment.
B—Recognition of hazardous thoughts.
C—Recognition of invulnerability in the situation.

Hazardous attitudes which contribute to poor pilot judgment can be effectively counteracted by redirecting that hazardous attitude so that appropriate action can be taken. Recognition of hazardous thoughts is the first step in neutralizing them in the ADM process. (PLT103) — AC 60-22

ALL
2195. What should a pilot do when recognizing a thought as hazardous?

A—Label that thought as hazardous, then correct that thought by stating the corresponding learned antidote.
B—Avoid developing this hazardous thought.
C—Develop this hazardous thought and follow through with modified action.

When a pilot recognizes a thought as hazardous, the pilot should label that thought as hazardous, then correct that thought by stating the corresponding antidote. (PLT103) — AC 60-22

ALL
2076. When a pilot recognizes a hazardous thought, he or she then should correct it by stating the corresponding antidote. Which of the following is the antidote for ANTIAUTHORITY?

A—It could happen to me.
B—Not so fast. Think first.
C—Follow the rules. They are usually right.

The antiauthority (don't tell me!) attitude is found in people who do not like anyone telling them what to do. The antidote for this attitude is: follow the rules, they are usually right. (PLT103) — AC 60-22

Answer (A) is incorrect because this is the invulnerability attitude. Answer (B) is incorrect because this is the impulsivity attitude.

Answers

2086 [A]	2073 [A]	2085 [B]	2074 [C]	2075 [B]	2195 [A]
2076 [C]					

ALL
2077. What is the antidote when a pilot has the hazardous attitude of 'Invulnerability'?

A—It can not be that bad.
B—It could happen to me.
C—It will not happen to me.

Invulnerability (it won't happen to me) is found in people who think accidents happen to others, but never to them. Pilots who think this way are more likely to take chances and increase risk. The antidote for this attitude is: It could happen to me. (PLT103) — AC 60-22

ALL
2176. Pilots should be familiar with positive alternatives to

A—Counteract hazardous attitudes.
B—Ensure the right choices have been made.
C—Ensure they are applying the appropriate antidote.

The pilot must examine decisions carefully to ensure choices have not been influenced by hazardous attitudes and be familiar with positive alternatives to counteract the hazardous attitudes (referred to as antidotes). (PLT103) — FAA-H-8083-25

ALL
2079. Aeronautical Decision Making (ADM) is a

A—mental process of analyzing all information in a particular situation and making a timely decision on what action to take.
B—systematic approach to the mental process used by pilots to consistently determine the best course of action for a given set of circumstances.
C—decision making process which relies on good judgment to reduce risks associated with each flight.

ADM is a systematic approach to the mental process used by aircraft pilots to consistently determine the best course of action in response to a given set of circumstances. (PLT022) — AC 60-22

ALL
2080. What is it often called when a pilot pushes his or her capabilities and the aircraft's limits by trying to maintain visual contact with the terrain in low visibility and ceiling?

A—Peer pressure.
B—Scud running.
C—Mind set.

Scud running is pushing the capabilities of the pilot and the aircraft to the limits by trying to maintain visual contact with the terrain while trying to avoid physical contact with it. (PLT022) — AC 60-22

Answer (A) is incorrect because peer pressure is poor decision making based upon emotional response to peers rather than evaluating a situation objectively. Answer (C) is incorrect because mind set is the inability to recognize and cope with changes in the situation different from those anticipated or planned.

ALL
2083. Ignoring minimum fuel reserve requirements is generally the result of overconfidence, disregarding applicable regulations, or

A—lack of flight planning.
B—impulsivity.
C—physical stress.

Ignoring minimum fuel reserve requirements is generally the result of overconfidence, lack of flight planning, or disregarding applicable regulations. (PLT104) — FAA-H-8083-25

ALL
2196. The basic drive for a pilot to demonstrate the "right stuff" can have an adverse effect on safety, by

A—a total disregard for any alternative course of action.
B—generating tendencies that lead to practices that are dangerous, often illegal, and may lead to a mishap.
C—allowing events, or the situation, to control his or her actions.

Pilots, particularly those with considerable experience, as a rule always try to complete a flight as planned, please passengers, meet schedules, and generally demonstrate that they have the "right stuff." These tendencies ultimately may lead to practices that are dangerous and often illegal, and may lead to a mishap. (PLT103) — AC 60-22

Answers

2077 [B]	2176 [A]	2079 [B]	2080 [B]	2083 [A]	2196 [B]

Risk Management

Risk Management is the part of the decision making process which relies on situational awareness, problem recognition, and good judgment to reduce risks associated with each flight. During each flight, decisions must be made regarding events involving interactions between the four risk elements—the pilot in command, the airplane, the environment, and the operation. The decision-making process involves an evaluation of each of these risk elements to achieve an accurate perception of the flight situation.

ALL

2084. Risk management, as part of the Aeronautical Decision Making (ADM) process, relies on which features to reduce the risks associated with each flight?

A—The mental process of analyzing all information in a particular situation and making a timely decision on what action to take.
B—Situational awareness, problem recognition, and good judgment.
C—Application of stress management and risk element procedures.

Risk Management is the part of the decision making process which relies on situational awareness, problem recognition, and good judgment to reduce risks associated with each flight. (PLT104) — AC 60-22

Checklists and Controls Exchange

Unjustified, sole reliance on the pilot's short- and long-term memory, using only regular flying skills on repetitive and too-familiar routes usually results in neglect of flight planning, preflight inspections, and checklists.

Checklists have been the foundation of pilot standardization and cockpit safety for years. The checklist is an aid to the memory and helps to ensure that critical items necessary for the safe operation of aircraft are not overlooked or forgotten. However, checklists are of no value if the pilot is not committed to their use. Without discipline and dedication to the use of checklists at the appropriate times, the odds are on the side of error. Pilots who fail to take the checklist seriously become complacent and the only thing they can rely on is memory.

The importance of consistent use of checklists cannot be overstated in pilot training. A major objective in primary flight training is to establish habit patterns that will serve pilots well throughout their entire flying career. The flight instructor must promote a positive attitude toward the use of checklists, and the student pilot must understand why checklists are important.

During flight training, there must always be a clear understanding between the student and flight instructor of who has control of the aircraft. Prior to any dual training flight, a briefing should be conducted that includes the procedure for the exchange of flight controls. The following positive, three-step process for the exchange of flight controls is highly recommended.

When a flight instructor wishes the student to take control of the aircraft, he/she should say to the student, "You have the flight controls." The student should acknowledge immediately by saying, "I have the flight controls." The flight instructor confirms by again saying, "You have the flight controls."

Answers
2084 [B]

Part of the procedure should be a visual check to ensure that the other person actually has the flight controls. When returning the controls to the flight instructor, the student should follow the same procedure the instructor used when giving control to the student. The student should stay on the controls until the instructor says, "I have the flight controls." There should never be any doubt as to who is flying the airplane at any one time. Numerous accidents have occurred due to a lack of communication or misunderstanding as to who actually had control of the aircraft, particularly between students and flight instructors. Establishing the above procedure during initial training will ensure the formation of a very beneficial habit pattern.

ALL

2081. What is one of the neglected items when a pilot relies on short and long term memory for repetitive tasks?

A—Flying outside the envelope.
B—Checklists.
C—Situation awareness.

Unjustified reliance on the pilot's short and long term memory, regular flying skills, repetitive and familiar routes usually results in neglect of flight planning, preflight inspections, and checklists. (PLT104) — AC 60-22

ALL

2197. A pilot who relies on short- and long-term memory for repetitive tasks often neglects

A—flying outside the envelope.
B—checklists.
C—situation awareness.

Unjustified reliance on the pilot's short- and long-term memory, regular flying skills, repetitive and familiar routes usually results in neglect of flight planning, preflight inspections, and checklists. (PLT104) — FAA-H-8083-25

ALL

2063. Consistent adherence to approved checklists is a sign of a

A—disciplined and competent pilot.
B—pilot who lacks the required knowledge.
C—low-time pilot.

To avoid missing important steps, always use the appropriate checklists whenever they are available. Consistent adherence to approved checklists is a sign of a disciplined and competent pilot. (PLT436) — FAA-H-8083-25

ALL

2068. To avoid missing important steps, always use the

A—appropriate checklists.
B—placarded airspeeds.
C—airworthiness certificate.

To avoid missing important steps, always use the appropriate checklists whenever they are available. Consistent adherence to approved checklists is a sign of a disciplined and competent pilot. (PLT122) — FAA-H-8083-25

ALL

2069. The positive three-step process in the exchange of flight controls between pilots includes these verbal steps: (1) You have the flight controls, (2) I have the flight controls and (3)

A—You have the flight controls.
B—I have the aircraft.
C—I have the flight controls.

There must always be a clear understanding who has control of the aircraft. Prior to flight, a briefing should be conducted that includes the procedure for the exchange of flight controls. A positive three-step process in the exchange of flight controls between pilots is a proven procedure: You have the flight controls, I have the flight controls, You have the flight controls. (PLT340) — FAA-H-8083-9

Answers

2081 [B]	2197 [B]	2063 [A]	2068 [A]	2069 [A]

Situational Awareness

Situational awareness is the accurate perception of the operational and environmental factors that affect the airplane, pilot, and passengers during a specific period of time. Maintaining situational awareness requires an understanding of the relative significance of these factors and their future impact on the flight. When situationally aware, the pilot has an overview of the total operation and is not fixated on one perceived significant factor. Some of the elements inside the airplane to be considered are the status of airplane systems, and also the pilot and passengers. In addition, an awareness of the environmental conditions of the flight, such as spatial orientation of the airplane, and its relationship to terrain, traffic, weather, and airspace must be maintained.

To maintain situational awareness, all of the skills involved in aeronautical decision making are used. For example, an accurate perception of pilot fitness can be achieved through self-assessment and recognition of hazardous attitudes. A clear assessment of the status of navigation equipment can be obtained through workload management, and establishing a productive relationship with ATC can be accomplished by effective resource use.

ALL

2082. An extreme case of a pilot getting behind the aircraft can lead to the operational pitfall of

A—loss of situational awareness.
B—loss of workload.
C—internal stress.

Getting behind the aircraft can result in not knowing where you are, an inability to recognize deteriorating circumstances, and/or the misjudgment of the rate of deterioration. (PLT104) — FAA-H-8083-25

Answers

2082 [A]

Chapter 4
Aircraft

Basic Aerodynamics

Four aerodynamic forces are considered to be basic because they act upon an aircraft during all flight maneuvers: there is the downward-acting force called **weight** which must be overcome by the upward-acting force called **lift**; and there is the rearward-acting force called **drag**, which must be overcome by the forward-acting force called **thrust**. During unaccelerated (straight-and-level) flight, the four aerodynamic forces which act on an aircraft are said to be in equilibrium, or: Lift = Weight and Thrust = Drag. As the velocity of a fluid (gas or liquid) increases, its pressure decreases. This is known as **Bernoulli's Principle**. For every action there is an equal and opposite reaction. This is known as **Newton's Third Law of Motion**. Both of these principles contribute to an airfoil's ability to generate lift. *See* Figure 4-1.

The speed for the **best glide** is one at which the airplane will travel the greatest forward distance for a given loss of altitude in still air. This best glide speed corresponds to an angle of attack resulting in the best lift-to-drag ratio (L/D_{MAX}).

FAA Figure 68 is a **VG diagram** that plots load factor against indicated airspeed and shows the pilot the limits within which the aircraft will safely handle structural loads. Point C is maneuvering speed (V_A). Any plotted combination of load factor and airspeed which falls in the shaded area may result in structural damage. V_{NO}, the maximum speed for normal operations, is shown by the vertical line from point D to point G on the VG diagram, and is marked as the upper limit of the green arc on the airspeed indicator. The red line on the airspeed indicator is represented by the line from point E to point F. The line connecting points C, D, and E represents the load factor above which structural damage may occur (the positive limit load factor). The VG diagram is applicable to any aircraft that uses an airfoil for its lift; however, it is least significant to PPCs because they typically operate at one speed in the "Normal Operating Range" of the diagram.

Ground effect occurs when flying within one wingspan or less above the surface. The airflow around the wing and wing tips is modified and the resulting pattern reduces the downwash and the induced drag. These changes can result in an aircraft becoming airborne before reaching recommended takeoff speed or floating during an approach to land. An airplane leaving ground effect after takeoff will require an increase in angle of attack to maintain the same lift coefficient, which in turn will cause an increase in induced drag and therefore, require increased thrust. Ground effect is negligible in a PPC because the wing is so high off the ground.

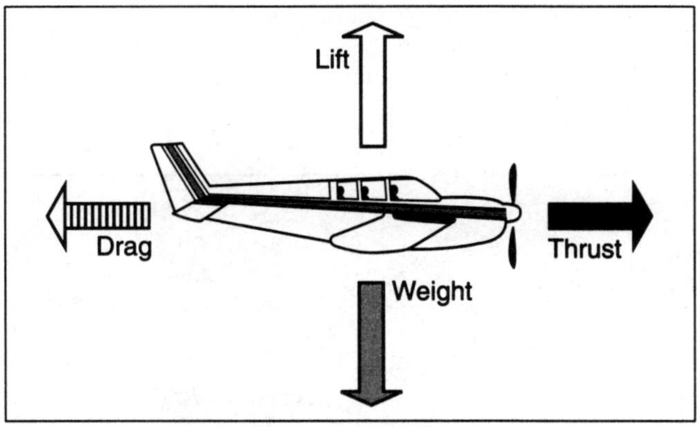

Figure 4-1. Relationship of forces in flight

Stability is the inherent ability of an aircraft to return, or not return, to its original flight condition after being disturbed by an outside force, such as rough air. An advantage of an aircraft said to be inherently stable is that it will require less effort to control. *See* Figure 4-2.

As the **angle of attack** is increased (to increase lift), the air will no longer flow smoothly over the upper wing surface but instead will become turbulent or "burble" near the trailing edge. A further increase in the angle of attack will cause the turbulent area to expand forward. At an angle of attack of approximately 18–20 degrees (for most wings), turbulence over the upper wing surface decreases lift so drastically that flight cannot be sustained and the wing stalls. The angle at which a stall occurs is called the critical angle of attack. An aircraft can stall at any airspeed or any attitude, but will always stall at the same critical angle of attack. *See* Figure 4-3.

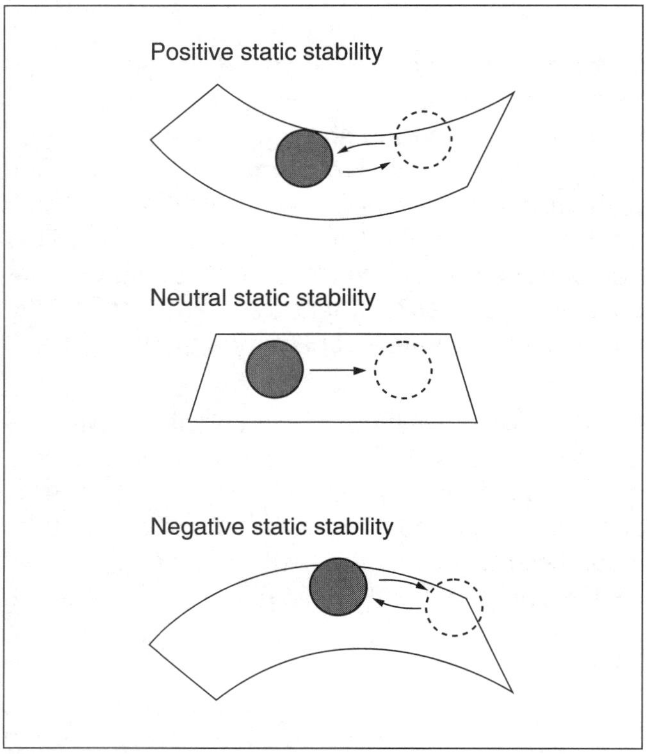

Figure 4-2. Static stability

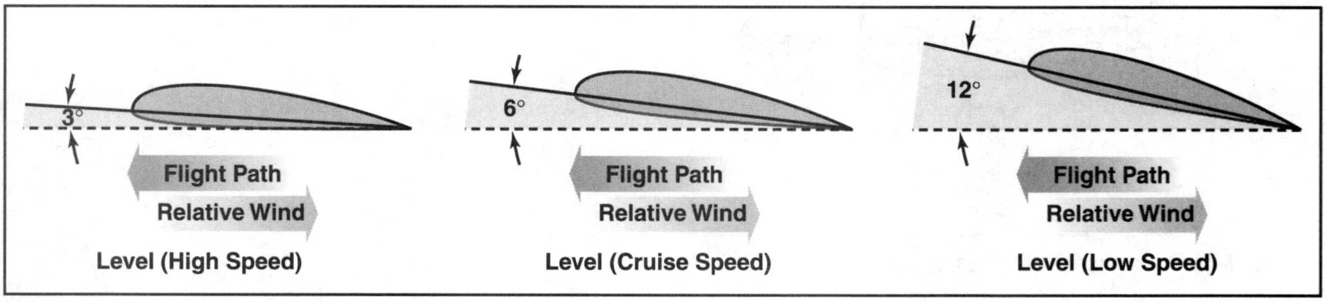

Figure 4-3. Angle of attack at various speeds

ALL

2225. The angle of attack at which an airfoil stalls will

A—increase if the CG is moved forward.
B—remain the same regardless of gross weight.
C—change with an increase in gross weight.

When the angle of attack is increased to between 18° and 20° (critical angle of attack) on most airfoils, the airstream can no longer follow the upper curvature of the wing because of the excessive change in direction. The airfoil will stall if the critical angle of attack is exceeded. The indicated airspeed at which stall occurs will be determined by weight and load factor, but the stall angle of attack is the same. (PLT477) — FAA-H-8083-25

Answers (A) and (C) are incorrect because an airfoil will always stall at the same angle of attack, regardless of the CG position or gross weight.

AIR, RTC, WSC, PPC

2237. What is the relationship of lift, drag, thrust, and weight when the airplane is in straight-and-level flight?

A—Lift equals weight and thrust equals drag.
B—Lift, drag, and weight equal thrust.
C—Lift and weight equal thrust and drag.

Lift and thrust are considered positive forces while weight and drag are considered negative forces, and the sum of the opposing forces is zero. That is, lift = weight, and thrust = drag. (PLT241) — FAA-H-8083-25

AIR, GLI, WSC

2239. The best speed to use for a glide is one that will result in the greatest glide distance for a given amount of

A—altitude.
B—fuel.
C—drag.

The best speed for the glide is one at which the airplane will travel the greatest forward distance for a given loss of altitude in still air. This best glide speed corresponds to an angle of attack resulting in the least drag on the airplane and giving the best lift-to-drag ratio (L/D_{MAX}). (PLT257) — FAA-H-8083-3

AIR, RTC, WSC, PPC

2229. Climb performance depends upon the

A—reserve power or thrust.
B—maximum L/D ratio.
C—cruise power setting.

Climb depends upon the reserve power or thrust. Reserve power is the available power over and above that required to maintain horizontal flight at a given speed. (PLT125) — FAA-H-8083-25

AIR, GLI, WSC

2215. (Refer to Figure 68.) The horizontal dashed line from point C to point E represents the

A—ultimate load factor.
B—positive limit load factor.
C—airspeed range for normal operations.

C to E is the maximum positive load limit. In this case it is 3.8 Gs, which is appropriate for normal category airplanes. (PLT074) — FAA-H-8083-25

Answer (A) is incorrect because "ultimate load factor" is not a real term; it is not depicted in the figure. Answer (C) is incorrect because this is depicted by the vertical line from point A to point J, to the vertical line from point D to point G.

AIR, GLI, WSC

2216. (Refer to Figure 68.) The vertical dashed line from point E to point F is represented on the airspeed indicator by the

A—upper limit of the yellow arc.
B—upper limit of the green arc.
C—blue radial line.

V_{NE} (never exceed airspeed), the vertical line from point E to F, is marked on airspeed indicators with a red radial line, the upper limit of the yellow arc. (PLT074) — FAA-H-8083-25

Answers

2225 [B]	2237 [A]	2239 [A]	2229 [A]	2215 [B]	2216 [A]

AIR, RTC, GLI, WSC

2223. What must a pilot be aware of as a result of ground effect?

A—Wingtip vortices increase creating wake turbulence problems for arriving and departing aircraft.
B—Induced drag decreases; therefore, any excess speed at the point of flare may cause considerable floating.
C—A full stall landing will require less up elevator deflection than would a full stall when done free of ground effect.

The reduction of the wing-tip vortices, due to ground effect, alters the spanwise lift distribution and reduces the induced angle of attack, and induced drag causing floating. (PLT131) — FAA-H-8083-25

Answer (A) is incorrect because wing-tip vortices are decreased. Answer (C) is incorrect because a full stall landing will require more up-elevator deflection, due to the increased lift in ground effect.

AIR, RTC, GLI, WSC, PPC

2224. An airplane said to be inherently stable will

A—be difficult to stall.
B—require less effort to control.
C—not spin.

A stable aircraft will tend to return to the original condition of flight if disturbed by a force such as turbulent air. This means that a stable airplane is easy to fly. (PLT213) — FAA-H-8083-25

Answer (A) is incorrect because stability of an aircraft has an effect on its stall recovery, not the difficulty of stall entry. Answer (C) is incorrect because an inherently stable aircraft can still spin.

AIR, RTC, WSC

2230. Maximum endurance is obtained at the point of minimum power to maintain the aircraft

A—in steady, level flight.
B—in a long range descent.
C—at its slowest possible indicated airspeed.

The maximum endurance condition is obtained at the point of minimum power required since this would require the lowest fuel flow to keep the airplane in steady, level flight. Maximum range condition occurs where the proportion between speed and power required is greatest. (PLT015) — FAA-H-8083-25

AIR, RTC, WSC

2236. When range and economy of operation are the principal goals, the pilot must ensure that the airplane will be operated at the recommended

A—specific endurance.
B—long-range cruise performance.
C—equivalent airspeed.

Total range is dependent on both fuel available and specific range. When range and economy of operation are the principal goals, the pilot must ensure that the airplane is operated at the recommended long-range cruise condition. By this procedure, the airplane will be capable of its maximum design-operating radius, or can achieve lesser flight distances with a maximum of fuel reserve at the destination. (PLT501) — FAA-H-8083-25

Answers

2223 [B]	2224 [B]	2230 [A]	2236 [B]

Pitot-Static Instruments

The pressure altimeter, vertical-speed indicator, and airspeed indicator operate in response to pressures through the **pitot-static system**. *See* Figure 4-4.

Static (atmospheric) **pressure** is taken from the static vents and is provided to all three instruments. Clogging of the static vents or line will cause all three instruments to become inoperative or to display erroneous readings.

Impact (ram) **pressure** is taken from the pitot tube and furnished to the airspeed indicator only. Clogging of the pitot opening will not affect operation of the altimeter or vertical speed indicator.

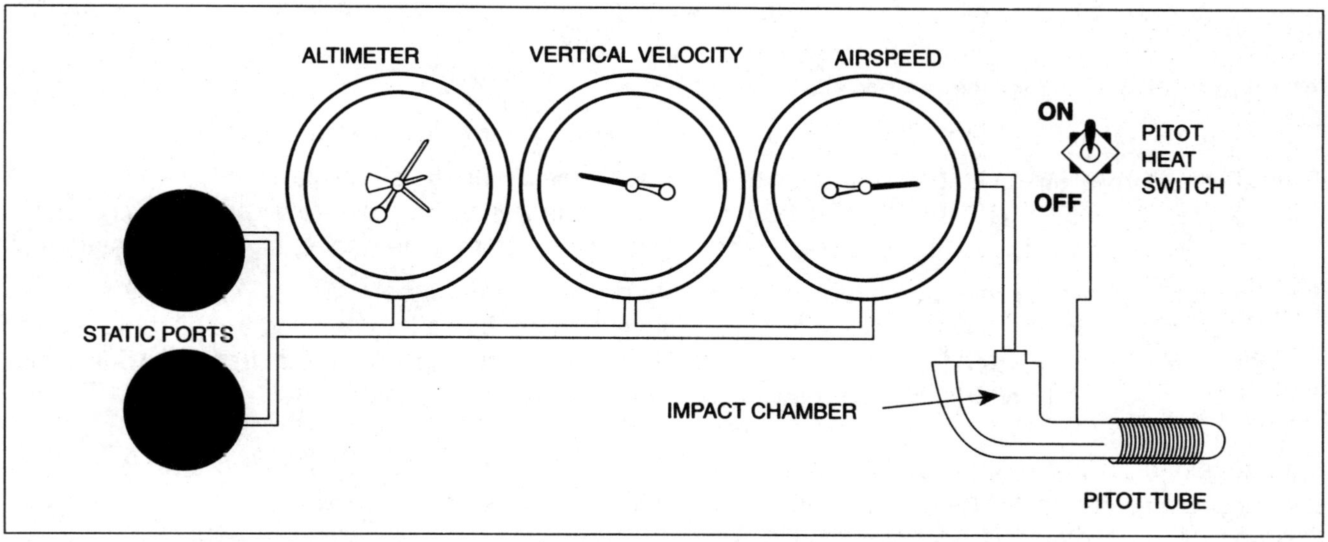

Figure 4-4. Pitot-static system

AIR, RTC, GLI, WSC

2339. The pitot system provides impact pressure for which instrument?

A—Altimeter.
B—Vertical-speed indicator.
C—Airspeed indicator.

The pitot tube provides input for the airspeed indicator only. (PLT337) — FAA-H-8083-25

Answers (A) and (B) are incorrect because the altimeter and vertical speed indicator operate off the static system.

AIR, RTC, GLI, WSC

2340. If the pitot tube and outside static vents become clogged, which instruments would be affected?

A—The altimeter, airspeed indicator, and turn-and-slip indicator.
B—The altimeter, airspeed indicator, and vertical speed indicator.
C—The altimeter, attitude indicator, and turn-and-slip indicator.

Airspeed, altimeter and vertical speed all receive static input and would indicate inaccurately if the static sources became plugged. (PLT337) — FAA-H-8083-25

Answer (A) is incorrect because the turn-and-slip indicator and attitude indicator are gyroscopic instruments, and are not part of the pitot-static system. Answer (C) is incorrect because the turn-and-slip indicator and attitude indicator are gyroscopic instruments, and are not part of the pitot-static system.

Answers
2339 [C] 2340 [B]

The Altimeter and Altitudes

An **altimeter** is an instrument used to measure height (altitude) by responding to atmospheric pressure changes. *See* Figure 4-5.

Altitude is indicated by three hands on the face of the altimeter. The shortest hand indicates altitude in tens of thousands of feet; the intermediate hand, in thousands of feet; and the longest hand in hundreds of feet. The altimeter is subdivided into 20-foot increments.

"**Altitude**" means elevation with respect to any assumed reference level, and different terms identify the reference level used. *See* Figure 4-6.

Indicated altitude—the altitude read on the altimeter after it is set to the current local altimeter setting.

Absolute altitude—the height above the surface.

True altitude—the true height above Mean Sea Level (MSL) normally measured in feet.

Pressure altitude—the altitude that is indicated whenever the altimeter setting dial (Kohlsman window) is adjusted to 29.92. This is the Standard Datum Plane; a theoretical level where air pressure is equal to 29.92 inches of mercury (in. Hg). The Standard Datum Plane may be above, at, or below sea level.

Density altitude—the pressure altitude corrected for nonstandard temperature and/or pressure. Rotating the setting knob on the altimeter simultaneously rotates the setting dial and the altimeter hands at a rate of one inch per 1,000 feet of altitude. Thus, increasing the setting dial from 29.15 to 29.85 would cause the hands of the altimeter to show an increase of 700 feet.

Prior to takeoff, the altimeter should be set to the current local altimeter setting. This is the value to which the scale of the altimeter is set so that the altimeter indicates true altitude at field elevation. If the altimeter setting is not available, the altimeter should be set to the elevation of the departure airport.

After takeoff, the altimeter should remain set to the current local altimeter setting until climbing through 18,000 feet MSL. At that time, the altimeter should be set to 29.92.

On a standard day (29.92" Hg and +15°C) at sea level, pressure altitude, true altitude, indicated altitude, and density altitude are all equal. Any variation from standard temperature or pressure will have an effect on the altimeter.

To compensate for the effect of nonstandard conditions, the altimeter must be set to the altimeter setting of a station within 100 NM of the aircraft (unless it is above 18,000 feet MSL).

If a flight is made from an area of low pressure/low temperature to an area of high pressure/high temperature without adjusting the altimeter setting, the altimeter will indicate *lower* than the actual altitude above ground level. If a flight is made from an area of high pressure/high temperature to an area of low pressure/low temperature without adjusting the altimeter setting, the altimeter will indicate *higher* than the actual altitude above mean sea level.

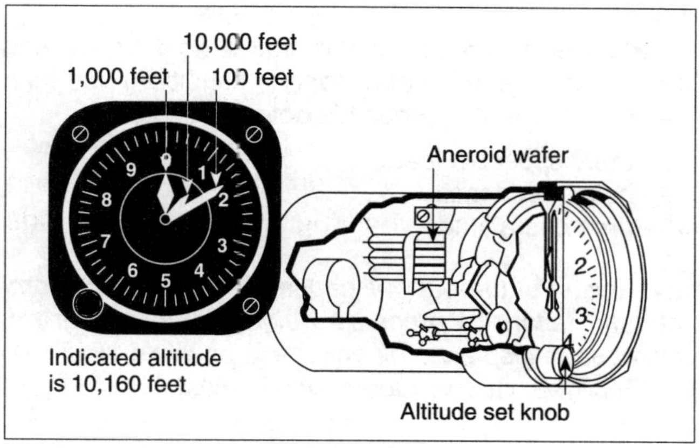

Figure 4-5. Altimeter components

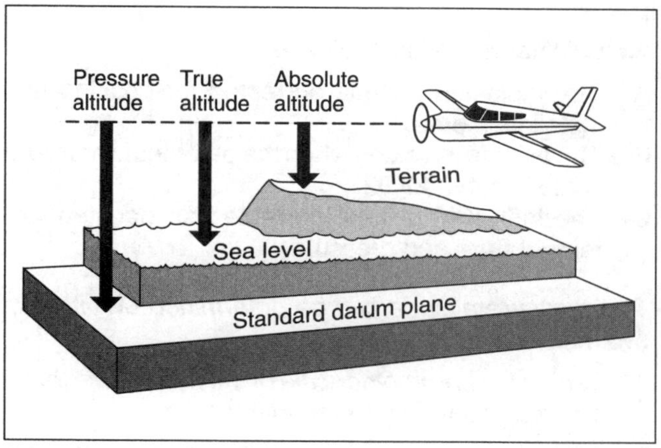

Figure 4-6. Types of altitude

ALL

2286. If it is necessary to set the altimeter from 29.15 to 29.85, what change occurs?

A—70-foot increase in indicated altitude.
B—70-foot increase in density altitude.
C—700-foot increase in indicated altitude.

When the knob on the altimeter is rotated, the pressure scale moves simultaneously with the altimeter pointers. The numerical values of pressure indicated in the window increase while the altimeter indicates an increase in altitude; or decrease while the altimeter indicates a decrease in altitude. This is contrary to the reaction on the pointers when air pressure changes, and is based solely on the mechanical makeup of the altimeter. The difference between the two settings is equal to 0.70" Hg (29.85 – 29.15). At the standard pressure lapse rate of 1" Hg = 1,000 feet in altitude, the amount of change equals 700 feet. (PLT305) — FAA-H-8083-25

ALL

2335. Under what condition is indicated altitude the same as true altitude?

A—If the altimeter has no mechanical error.
B—When at sea level under standard conditions.
C—When at 18,000 feet MSL with the altimeter set at 29.92.

On a standard day (29.92" Hg and +15°C) at sea level, pressure altitude, indicated altitude, and density altitude are all equal. Any variation from standard temperature or pressure will have an effect on the altimeter. (PLT023) — FAA-H-8083-25

Answer (A) is incorrect because mechanical error does not apply to true altitude. Answer (C) is incorrect because when the altimeter is set to 29.92, it indicates pressure altitude.

Answers
2286 [C] 2335 [B]

ALL

2336. What is pressure altitude?

A—The indicated altitude corrected for position and installation error.
B—The altitude indicated when the barometric pressure scale is set to 29.92.
C—The indicated altitude corrected for nonstandard temperature and pressure.

The pressure altitude can be determined by either of two methods:

1. *Setting the barometric scale of the altimeter to 29.92 and reading the indicated altitude, or*
2. *Applying a correction factor to the elevation (true altitude) according to the reported altimeter setting.*

(PLT023) — FAA-H-8083-25

Answer (A) is incorrect because the altimeter is not corrected for position and installation error. Answer (C) is incorrect because indicated altitude corrected for nonstandard temperature and pressure defines density altitude.

ALL

2337. What is true altitude?

A—The vertical distance of the aircraft above sea level.
B—The vertical distance of the aircraft above the surface.
C—The height above the standard datum plane.

True altitude is height above sea level. Airport terrain and obstacle elevations found on aeronautical charts are true altitudes. (PLT023) — FAA-H-8083-25

Answer (B) is incorrect because the vertical distance above the surface is absolute altitude. Answer (C) is incorrect because the height above the standard datum plane is pressure altitude.

LTA

2321. The maximum altitude that a rigid airship can reach (under a given atmospheric condition) and then return safely to the surface is determined by

A—the disposable load.
B—ballonet capacity.
C—pressure altitude.

Expansion of the gas will cause the ballonets to completely deflate at pressure height. Thus, no further increase in displaced air volume is possible. (PLT152) — Goodyear Airship Operations Manual

Answers

2336 [B] 2337 [A] 2321 [C]

Airspeeds and the Airspeed Indicator

A pilot must be familiar with the following airspeed terms and abbreviations:

Indicated Airspeed (IAS)—the uncorrected reading obtained from the airspeed indicator.

Calibrated Airspeed (CAS)—indicated airspeed corrected for installation and instrument error.

True Airspeed (TAS)—calibrated airspeed corrected for temperature and pressure variations.

A number of airspeed limitations, abbreviated as "V" speeds, are indicated by color-coded marking on the airspeed indicator (*See* Figure 4-7):

V_{SO}—stall speed or minimum steady flight speed in the landing configuration (the lower limit of the white arc).

V_{FE}—maximum flap extended speed (the upper limit of the white arc). The entire white arc defines the flap operating range.

V_{S1}—the stall speed or minimum steady flight speed in a specified configuration (the lower limit of the green arc). The entire green arc defines the normal operating range.

V_{NO}—the maximum structural cruising speed (the upper limit of the green arc and lower limit of the yellow arc). The yellow arc defines the caution range, which should be avoided unless in smooth air.

V_{NE}—never exceed speed (the upper limit of the yellow arc) marked in red.

There are other important airspeed limitations that are not color-coded on the airspeed indicator:

V_{LE}—the maximum landing gear extended speed.

V_A—the design maneuvering speed. If rough air or severe turbulence is encountered, airspeed should be reduced to maneuvering speed or less to minimize stress on the airplane structure.

V_Y—the best rate-of-climb speed (the airspeed that will result in the most altitude in a given period of time).

V_X—the best angle-of-climb speed (the airspeed that will result in the most altitude in a given distance).

AIR, RTC, GLI, WSC

2334. What does the red line on an airspeed indicator represent?

A—Maneuvering speed.
B—Turbulent or rough-air speed.
C—Never-exceed speed.

The upper end of the arc is marked by a red radial line which is the never-exceed speed (V_{NE}). (PLT132) — FAA-H-8083-25

Answer (A) is incorrect because the maneuvering speed and turbulent or rough-air speed is not indicated on the airspeed indicator. Answer (B) is incorrect because the maneuvering speed and turbulent or rough-air speed is not indicated on the airspeed indicator.

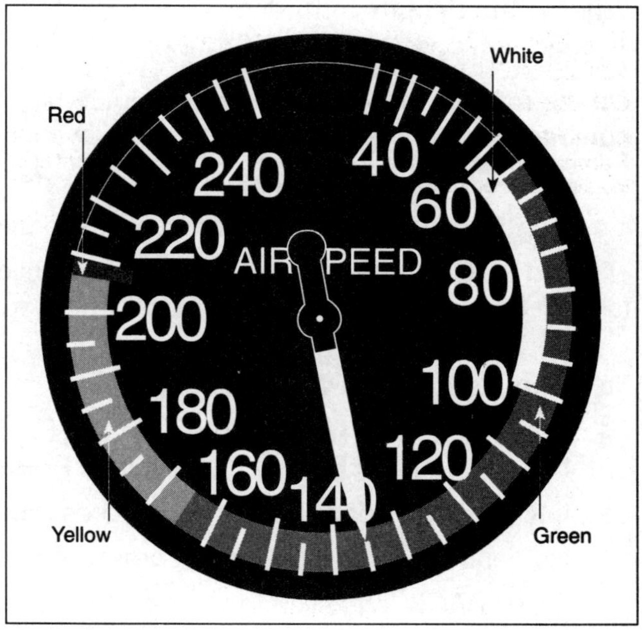

Figure 4-7. Airspeed indicator

Answers
2334 [C]

Density Altitude and Aircraft Performance

Density altitude is the pressure altitude corrected for nonstandard temperature and/or pressure. The effect of pressure altitude and ambient temperature is to define density altitude and its effect on landing performance. On a hot day, the air becomes "thinner" or lighter, and its density is equivalent to a higher altitude in the standard atmosphere, thus the term "high density altitude."

Aircraft performance charts show a pilot what can be expected of an aircraft (rate of climb, takeoff roll, etc.) under stipulated conditions. Prediction of performance is based upon a sea level temperature of +15°C (+59°F) and atmospheric pressure of 29.92" Hg (1013.2 mb). This combination of temperature and pressure is called a "standard day." When the air is at a "standard density," temperature and/or pressure deviations from standard will change the air density, or the density altitude, which affects aircraft performance. Performance charts allow the pilot to predict how an aircraft will perform.

Relative humidity also affects density altitude, but is not considered when the performance charts are formulated. A combination of high temperature, high humidity, and high altitude result in a density altitude higher than the pressure altitude which, in turn, results in reduced aircraft performance.

High density altitude results in a faster takeoff speed, longer takeoff roll, less climb rate, less ability to get over a 50-foot obstacle, lower service ceiling, less angle of climb, faster landing speed, and longer landing distance. Comparing an 8,000-foot density altitude to standard sea level conditions, it takes about twice as far to take off in the high density altitude conditions.

High density altitude problem:

Altimeter setting.................................... 29.0

Airport temperature ....95°F

Airport altitude ....3,894 feet MSL

What is the density altitude?

To solve this using FAA Figure 24:

On the right-hand side, altimeter setting 29.0" Hg is 863 feet. Since it is a low pressure and a positive number, it must be added to the airport elevation:

3,894 + 863 is 4,757 pressure altitude corrected for barometric pressure.

The 95°F must be converted to Celsius to use this chart. To convert, use FAA Figure 2—35°C.

Find 35°C on the bottom of the graph and follow it up until you hit the pressure altitude of 4,757; then follow it horizontally to 8,000-foot density altitude.

To calculate it on the CX-2 calculator:

1. Press the **FLIGHT** button to get to the **FLIGHT** menu, and press **1** or **ENT** to select **1. Altitude**; then press **1** or **ENT** again to select **1. Press Alt** (pressure altitude).
2. **Ialt** input **3894** (Indicated altitude), press **ENT**.
3. Hg" input 29, press **ENT** to get pressure altitude as an answer.
4. Press **BACK** to back up one step in the menu.
5. Press 2 to select **2. Density Alt**, or arrow down to **2. Density Alt** and press **ENT** to select.
6. The pressure altitude you just calculated appears at the top (the computer remembers it).

7. To convert 95°F to Celsius, arrow down or press **ENT** and input **95°F** in the temperature entry, and press the **second** function button and the **F to C** conversion key (number 3); 95°F is converted to 35°C.
8. With 35°C as the temperature entry, press **ENT** and 8,035 feet is displayed as the density altitude.

To use the E6-B circular slide rule calculator:

This can be done if you know the pressure altitude (this is typically done by setting the altimeter to 29.92 inches and reading the pressure altitude from the altimeter). First convert Fahrenheit to Celsius on the temperature conversion scale. Under the density altitude window, place the temperature over the pressure altitude and read off the density altitude in the density altitude window.

ALL

2217. What is absolute altitude?

A—The altitude read directly from the altimeter.
B—The vertical distance of the aircraft above the surface.
C—The height above the standard datum plane.

Absolute altitude is height above the surface. This height may be indicated directly on a radar altimeter (if so equipped). Absolute altitude may be approximately computed from indicated altitude and chart elevation data. (PLT023) — FAA-H-8083-25

Answer (A) is incorrect because the altitude read from the altimeter is indicated altitude. Answer (C) is incorrect because the height above the standard datum plane is pressure altitude.

ALL

2011. What is density altitude?

A—The height above the standard datum plane.
B—The pressure altitude corrected for nonstandard temperature.
C—The altitude read directly from the altimeter.

Under standard atmospheric conditions, each level of air in the atmosphere has a specific density, and under standard conditions, pressure altitude and density altitude identify the same level. At temperatures higher or lower than standard, density altitude cannot be determined directly from the altimeter. (PLT127) — FAA-H-8083-25

Answer (A) is incorrect because the height above the standard datum plane is pressure altitude. Answer (C) is incorrect because the altitude read from the altimeter is indicated altitude.

ALL

2067. Density altitude, and its effect on landing performance, is defined by

A—pressure altitude and ambient temperature.
B—headwind and landing weight.
C—humidity and braking friction forces.

The effect of pressure altitude and ambient temperature is to define density altitude and its effect on landing performance. (PLT127) — FAA-H-8083-25

ALL

2227. What effect does high density altitude, as compared to low density altitude, have on propeller efficiency and why?

A—Efficiency is increased due to less friction on the propeller blades.
B—Efficiency is reduced because the propeller exerts less force at high density altitudes than at low density altitudes.
C—Efficiency is reduced due to the increased force of the propeller in the thinner air.

The propeller produces thrust in proportion to the mass of air being accelerated through the rotating blades. If the air is less dense, propeller efficiency is decreased. (PLT351) — FAA-H-8083-25

Answers

2217 [B]	2011 [B]	2067 [A]	2227 [B]

ALL

2355. (Refer to Figure 24.) Determine the pressure altitude at an airport that is 1,386 feet MSL with an altimeter setting of 29.97

A—1,341 feet MSL.
B—1,451 feet MSL
C—1,562 feet MSL.

Referencing FAA Figure 24, use the following steps:

1. *Since the given altimeter setting is not shown in FAA Figure 24, interpolation is necessary. Locate the settings immediately above and below the given value of 29.97" Hg:*

Altimeter Setting	*Conversion Factor*
29.92	*0 feet*
30.00	*-73 feet*

2. *Determine the difference between the two conversion factors: 0 – 73 = -73 feet*
3. *Determine the amount of the difference to be subtracted from the 29.92" Hg conversion factor:*

   (-73 ÷ 8) x 5 = -45.6 or 46 feet
4. *Subtract the conversion factor from the airport elevation to determine the pressure altitude:*

   1,386 – 46 = 1,340 feet MSL (pressure altitude)

(PLT041) — FAA-H-8083-25

ALL

2356. (Refer to Figure 24.) What is the effect of a temperature increase from 30 to 50°F on the density altitude if the pressure altitude remains at 3,000 feet MSL?

A—900-foot increase.
B—1,100-foot decrease.
C—1,300-foot increase.

Referencing FAA Figure 24, use the following steps:

1. *Enter the density altitude chart at 30°F. Proceed upward to the 3,000-foot pressure altitude line. From the point of intersection, move left to the edge of the chart and read a density altitude of 1,650 feet.*
2. *Enter the density altitude chart at 50°F. Proceed upward to the 3,000-foot pressure altitude line. From the point of intersection, move left to the edge of the chart and read a density altitude of 3,000 feet.*
3. *Find the difference between the two values: 3,000 – 1,650 = 1,350 foot (increase)*

(PLT005) — FAA-H-8083-25

ALL

2008. What effect, if any, does high humidity have on aircraft performance?

A—It increases performance.
B—It decreases performance.
C—It has no effect on performance.

An increase in air temperature or humidity, or a decrease in air pressure (which results in a higher density altitude) will significantly decrease both power output and propeller efficiency. If an air mass is humid, there is more water in it, therefore, less oxygen. (PLT124) — FAA-H-8083-25

ALL

2208. What are the standard temperature and pressure values for sea level?

A—15°C and 29.92" Hg.
B—59°C and 1013.2 millibars.
C—59°F and 29.92 millibars.

Standard sea level pressure is 29.92 inches of mercury. Standard sea level temperature is 15°C. (PLT512) — FAA-H-8083-25

ALL

2009. What effect does high density altitude have on aircraft performance?

A—It increases engine performance.
B—It reduces climb performance.
C—It increases takeoff performance.

An increase in air temperature or humidity, or a decrease in air pressure (which results in a higher density altitude) will significantly decrease both power output and propeller efficiency. (PLT124) — FAA-H-8083-25

ALL

2168. Which factor would tend to increase the density altitude at a given airport?

A—An increase in barometric pressure.
B—An increase in ambient temperature.
C—A decrease in relative humidity.

On a hot day, the air becomes "thinner" or lighter, and its density is equivalent to a higher altitude in the standard atmosphere, thus the term "high density altitude." (PLT206) — AC 00-6A, Chapter 3

Answer (A) is incorrect because barometric pressure increases do not directly affect density altitude changes. Answer (C) is incorrect because a decrease in relative humidity does not affect density altitude changes.

Answers

2355 [A]	2356 [C]	2008 [B]	2208 [A]	2009 [B]	2168 [B]

Takeoff and Landing Considerations

Operating with an uphill runway slope has no effect on the takeoff speed, but will increase the takeoff distance. The most critical conditions of takeoff performance are the result of some combination of high gross weight, altitude, temperature, and unfavorable wind. In all cases, the pilot must make an accurate prediction of takeoff distance from the performance data of the AFM/POH, regardless of the runway available, and strive for a polished, professional takeoff procedure.

ALL

2010. Which combination of atmospheric conditions will reduce aircraft takeoff and climb performance?

A—Low temperature, low relative humidity, and low density altitude.
B—High temperature, low relative humidity, and low density altitude.
C—High temperature, high relative humidity, and high density altitude.

An increase in air temperature or humidity, or a decrease in air pressure (which results in a higher density altitude), will significantly decrease both power output and propeller efficiency. (PLT124) — FAA-H-8083-25

Answer (A) is incorrect because all of these conditions improve performance. Answer (B) is incorrect because low humidity and low density altitude improve performance.

ALL

2199. What effect does an uphill runway slope have on takeoff performance?

A—Increases takeoff speed.
B—Increases takeoff distance.
C—Decreases takeoff distance.

The effect of runway slope on takeoff distance is due to the component of weight along the inclined path of the aircraft. An upslope would contribute a retarding force component, while a downslope would contribute an accelerating force component. In the case of an upslope, the retarding force component adds to drag and rolling friction to reduce the net accelerating force. (PLT129) — FAA-H-8083-25

Answer (A) is incorrect because runway slope has no effect on the takeoff speed. Answer (C) is incorrect because an upslope will increase the takeoff distance required.

AIR, RTC, GLI, WSC, PPC

2283. The most critical conditions of takeoff performance are the result of some combination of high gross weight, altitude, temperature, and

A—unfavorable wind.
B—obstacles surrounding the runway.
C—powerplant systems.

The most critical conditions of takeoff performance are the result of some combination of high gross weight, altitude, temperature, and unfavorable wind. In all cases, the pilot must make an accurate prediction of takeoff distance from the performance data of the AFM/POH, regardless of the runway available, and strive for a polished, professional takeoff procedure. (PLT134) — FAA-H-8083-25

Answers

2010 [C] 2199 [B] 2283 [A]

Carburetor Ice

As air flows through a carburetor it expands rapidly. At the same time, fuel forced into the airstream is vaporized. Expansion of the air and vaporization of the fuel causes a sudden cooling of the mixture which may cause ice to form inside the carburetor. The possibility of icing should always be considered when operating in conditions where the temperature is between 20°F and 70°F, and the relative humidity is high.

Carburetor heat preheats the air before it enters the carburetor and either prevents carburetor ice from forming or melts any ice which may have formed. When carburetor heat is applied, the heated air that enters the carburetor is less dense. This causes the fuel/air mixture to become enriched, and this in turn decreases engine output and increases engine operating temperatures.

During engine runup prior to departure from a high-altitude airport, the pilot may notice a slight engine roughness which is not affected by the magneto check, but grows worse during the carburetor heat check. In this case, the air/fuel mixture may be too rich due to the lower air density at the high altitude, and applying carburetor heat decreases the air density even more. A leaner setting of the mixture control may correct this problem.

In an airplane with a fixed-pitch propeller, the first indication of carburetor ice would likely be a decrease in RPM as the air supply is choked off. Application of carburetor heat will decrease air density, causing the RPM to drop even lower. Then, as the carburetor ice melts, the RPM will rise gradually.

Fuel injection systems do not utilize a carburetor and are generally considered to be less susceptible to icing than carburetor systems are.

AIR, RTC, WSC, PPC
2341. Which condition is most favorable to the development of carburetor icing?

A—Any temperature below freezing and a relative humidity of less than 50 percent.
B—Temperature between 32 and 50°F and low humidity.
C—Temperature between 20 and 70°F and high humidity.

If the temperature is between -7°C (20°F) and 21°C (70°F) with visible moisture or high humidity, the pilot should be constantly on the alert for carburetor ice. (PLT190) — FAA-H-8083-25

Answer (A) is incorrect because carburetor icing is more likely with high humidity. Answer (B) is incorrect because carburetor icing is more likely with high humidity.

Aviation Fuel

Fuel does two things for the engine; it acts both as an agent for combustion and as an agent for cooling (based on the mixture setting of the engine).

Aviation fuel is available in several grades. The proper grade for a specific engine will be listed in the aircraft flight manual. If the proper grade of fuel is not available, it is possible to use the next higher grade. A lower grade of fuel should never be used.

The use of low-grade fuel or an air/fuel mixture which is too lean may cause **detonation**, which is the uncontrolled spontaneous explosion of the mixture in the cylinder. Detonation produces extreme heat.

Preignition is the premature burning of the air/fuel mixture. It is caused by an incandescent area (such as a carbon or lead deposit heated to a red hot glow) which serves as an ignitor in advance of normal ignition.

Answers
2341 [C]

Fuel can be contaminated by water and/or dirt. The air inside the aircraft fuel tanks can cool at night, and this cooling forms water droplets (through condensation) on the insides of the fuel tanks. These droplets then fall into the fuel. To avoid this problem, always fill the tanks completely when parking overnight.

Thoroughly drain all of the aircraft's sumps, drains, and strainers before a flight to get rid of all the water that might have collected.

Dirt can get into the fuel if refueling equipment is poorly maintained or if the refueling operation is sloppy. Use care when refueling an aircraft.

Two fuel pump systems are used on most airplanes. The main pump system is engine driven and an auxiliary electric driven pump is provided for use in the event the engine pump fails. The auxiliary pump, commonly known as the "boost pump," provides added reliability to the fuel system, and is also used as an aid in engine starting. The electric auxiliary pump is controlled by a switch in the cockpit.

AIR, RTC, WSC, PPC

2342. On aircraft equipped with fuel pumps, when is the auxiliary electric driven pump used?

A—All the time to aid the engine-driven fuel pump.
B—In the event engine-driven fuel pump fails.
C—Constantly except in starting the engine.

Two fuel pump systems are used on most airplanes. The main pump system is engine driven and an auxiliary electric driven pump is provided for use in the event the engine pump fails. The auxiliary pump, commonly known as the "boost pump," provides added reliability to the fuel system, and is also used as an aid in engine starting. The electric auxiliary pump is controlled by a switch in the cockpit. (PLT253) — FAA-H-8083-25

AIR, RTC, WSC

2344. If a pilot suspects that the engine (with a fixed-pitch propeller) is detonating during climb-out after takeoff, the initial corrective action to take would be to

A—lean the mixture.
B—lower the nose slightly to increase airspeed.
C—apply carburetor heat.

To prevent detonation, the pilot should use the correct grade of fuel, maintain a sufficiently rich mixture, open the throttle smoothly, and keep the temperature of the engine within recommended operating limits. Some aircraft have an automatically enriched mixture for enhanced cooling in takeoff and climb-out at full throttle. Lowering the nose will allow the aircraft to gain airspeed, which eventually lowers the engine temperature. (PLT115) — FAA-H-8083-25

Answer (A) is incorrect because leaning the mixture increases engine temperatures; detonation results from excessively high engine temperatures. Answer (C) is incorrect because although a richer fuel mixture results from applying carburetor heat, the heat may offset the cooling effect of the mixture change. The most efficient initial action would be to increase airspeed.

AIR, RTC, WSC, PPC

2345. The uncontrolled firing of the fuel/air charge in advance of normal spark ignition is known as

A—combustion.
B—pre-ignition.
C—detonation.

Preignition is defined as ignition of the fuel prior to normal ignition. (PLT478) — FAA-H-8083-25

Answer (A) is incorrect because combustion is the normal engine process. Answer (C) is incorrect because detonation is the exploding of the fuel/air mixture.

Answers
2342 [B] 2344 [B] 2345 [B]

Weight and Balance

Even though an aircraft has been certificated for flight at a specified maximum gross weight, it may not be safe to take off with that load under all conditions. High altitude, high temperature, and high humidity are additional factors which may require limiting the load to some weight less than the maximum allowable.

Some of the problems caused by overloading an aircraft are:

- The aircraft will need a higher takeoff speed, which results in a longer takeoff run.
- Both the rate and angle of climb will be reduced.
- The service ceiling will be lowered.
- The cruising speed will be reduced.
- The cruising range will be shortened.
- Maneuverability will be decreased.
- A longer landing roll will be required because the landing speed will be higher.
- Excessive loads will be imposed on the structure, especially the landing gear.

The **empty weight** is obtained from appropriate charts. It includes the airframe, engine, all fixed equipment, and unusable fuel and oil. Some aircraft include all oil in the aircraft empty weight. The **useful load** includes the pilot, passengers, baggage, fuel and oil. The **takeoff weight** is the empty weight plus the useful load. The **landing weight** is the takeoff weight minus any fuel used.

The **arm** is the horizontal distance measured in inches from the datum line to a point on the aircraft. If measured aft, toward the tail, the arm is given a positive (+) value; if measured forward, toward the nose, the arm is given a negative (-) value. *See* Figure 4-8.

The **moment** is the product of the weight of an object multiplied by its arm and is expressed in pound-inches (lbs-in). A formula that is used to find moment is usually expressed as follows:

Weight × Arm = Moment

The **moment index** is a moment divided by a constant such as 100 or 1,000. It is used to simplify computations where heavy items and long arms result in large, unmanageable numbers. This is usually expressed as Moment/100 or Mom/1,000, etc.

The center of gravity (CG) is the point about which an aircraft will balance, and it is expressed in inches from datum. The center of gravity is found by dividing the total moment by the total weight, and the formula is usually expressed as follows:

$$\frac{\text{Total Moment}}{\text{Total Weight}} = \text{CG (inches aft of datum)}$$

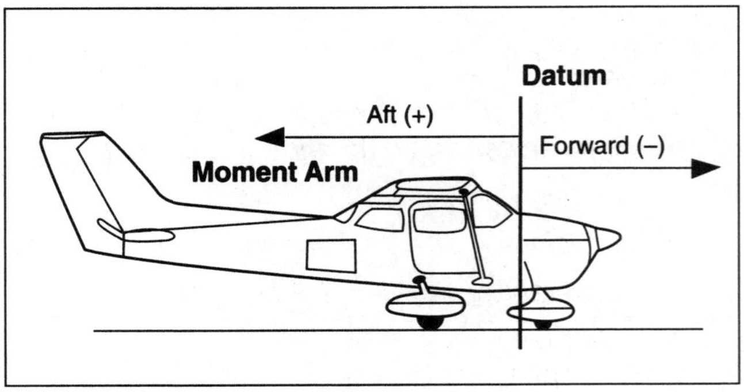

Figure 4-8. Positive and negative arm

Airplane

In addition to considering the weight to be carried, the pilot must ensure that the load is arranged to keep the aircraft in balance. The balance point, or Center of Gravity (CG), is the point at which all of the weight of the airplane is considered to be concentrated. For an aircraft to be safe to fly, the center of gravity must fall between specified limits. All aircraft categories have unique balance requirements; some need to be balanced precisely while others are less sensitive to fore and aft weight distribution. The specific balance requirements for each aircraft category are covered in the Pilot Operating Handbook (POH) and will be tested on during the practical (checkride). To keep the CG within safe limits it may be necessary to move weight toward the nose of the aircraft (forward), which moves the CG forward, or toward the tail (aft) which moves the CG aft.

The specified forward and aft points within which the CG must be located for safe flight are called the center of gravity limits. These limits are specified by the manufacturer. The distance between the forward and aft CG limits is called the center of gravity range.

Weight-Shift Control

Weight and loading is typically used for weight-shift control. The cart normally does not have balance requirements, but the attachment point to the cart is specified in the Pilot's Operating Handbook.

Powered Parachute

Weight and loading is used for powered parachutes. The POH procedure for the fore and aft attachment points of the wing to the cart must be followed so that the cart will obtain the proper deck angle (the angle of the cart bottom with the horizontal plane).

ALL
2174. Problems caused by overloading an aircraft include

A—reduced climb rate, excessive structural loads, and shortened cruising range.
B—increased service ceiling, increased angle of climb, and increased cruising speed.
C—slower takeoff speed, increased maneuverability, and shorter takeoff roll.

Some of the problems caused by overloading an aircraft are:

- *The aircraft will need a higher takeoff speed, which results in a longer takeoff run.*
- *Both the rate and angle of climb will be reduced.*
- *The service ceiling will be lowered.*
- *The cruising speed will be reduced.*
- *The cruising range will be shortened.*
- *Maneuverability will be decreased.*
- *A longer landing roll will be required because the landing speed will be higher.*
- *Excessive loads will be imposed on the structure, especially the landing gear.*

(PLT313) — FAA-H-8083-1

ALL
2182. Most aircraft are so designed that if all seats are occupied, all baggage allowed by the baggage compartment structure is carried, and all the fuel tanks are full, the aircraft will be

A—grossly overloaded.
B—under maximum gross weight.
C—at maximum gross weight.

The maximum allowable gross weight for an aircraft is determined by design considerations. However, the maximum operational weight may be less than the maximum allowable due to such considerations as high density altitude or high-drag field conditions caused by wet grass or water on the runway. The maximum gross weight may also be limited by the departure or arrival airport's runway length. One important preflight consideration is the distribution of the load in the aircraft. Loading an aircraft so the gross weight is less than the maximum allowable is not enough. This weight must be distributed to keep the CG within the limits specified in the POH or AFM. (PLT313) — FAA-H-8083-1

Answers
2174 [A] 2182 [A]

ALL

2183. When operating an aircraft, the pilot-in-command is responsible for using

A—the most current weight and balance data.
B—weight and balance data from the factory.
C—recent weight and balance data.

It is the responsibility of the pilot in command to use the most current weight and balance data when planning a flight and operating the aircraft. (PLT328) — FAA-H-8083-1

AIR, RTC

2350. (Refer to Center-of Gravity Envelope and Loading Graph figure on the next page.) GIVEN:

	Weight	Moment
Gyroplane basic weight	1,315	150.1 (oil included)
Pilot weight	140	?
Passenger weight	150	?
27 gal fuel	162	?

The CG is located

A—outside the CG envelope; the maximum gross weight is exceeded.
B—outside the CG envelope; the maximum gross weight and the gross-weight moment are exceeded.
C—within the CG envelope; neither maximum gross weight nor gross-weight moment is exceeded.

Total the weights and moments to determine if the CG is within limits.

Item	*Weight*	*Moment*
Gyroplane	*1,315*	*150.1*
Pilot	*140*	*7.2*
Passenger	*150*	*12.6*
27 gallons gas	*+ 162*	*+ 17.8*
Total	*1,767*	*187.7*

These figures fall within permitted weight and CG. (PLT021) — FAA-H-8083-1

AIR, RTC

2351. (Refer to Center-of Gravity Envelope and Loading Graph figure on the next page.) GIVEN:

	Weight	Moment
Gyroplane basic weight	1,315	154.0 (oil included)
Pilot weight	145	?
Passenger weight	153	?
27 gal fuel	162	?

The CG is located

A—outside the CG envelope; the maximum gross weight is exceeded.
B—outside the CG envelope; the maximum gross weight and the gross-weight moment are exceeded.
C—within the CG envelope; neither maximum gross weight nor gross-weight moment is exceeded.

Total the weights and moments to determine if the CG is within limits.

Item	*Weight*	*Moment*
Gyroplane	*1,315*	*154.0*
Pilot	*145*	*7.4*
Passenger	*153*	*12.8*
27 gallons gas	*+ 162*	*+ 17.6*
Total	*1,775*	*191.8*

These figures fall within permitted weight but outside CG. (PLT021) — FAA-H-8083-1

AIR, RTC, WSC, PPC

2323. Which items are included in the empty weight of an aircraft?

A—Unusable fuel and undrainable oil.
B—Only the airframe, powerplant, and optional equipment.
C—Full fuel tanks and engine oil to capacity.

Empty weight consists of the airframe, engines, and all items of operating equipment that have fixed locations and are permanently installed in the airplane. It includes optional and special equipment, fixed ballast, hydraulic fluid, unusable (residual) fuel, and undrainable (residual) oil. (PLT328) — FAA-H-8083-25

Answer (B) is incorrect because empty weight also includes operating fluids and unusable fuel. Answer (C) is incorrect because empty weight does not include full fuel.

Answers

2183 [A] 2350 [C] 2351 [B] 2323 [A]

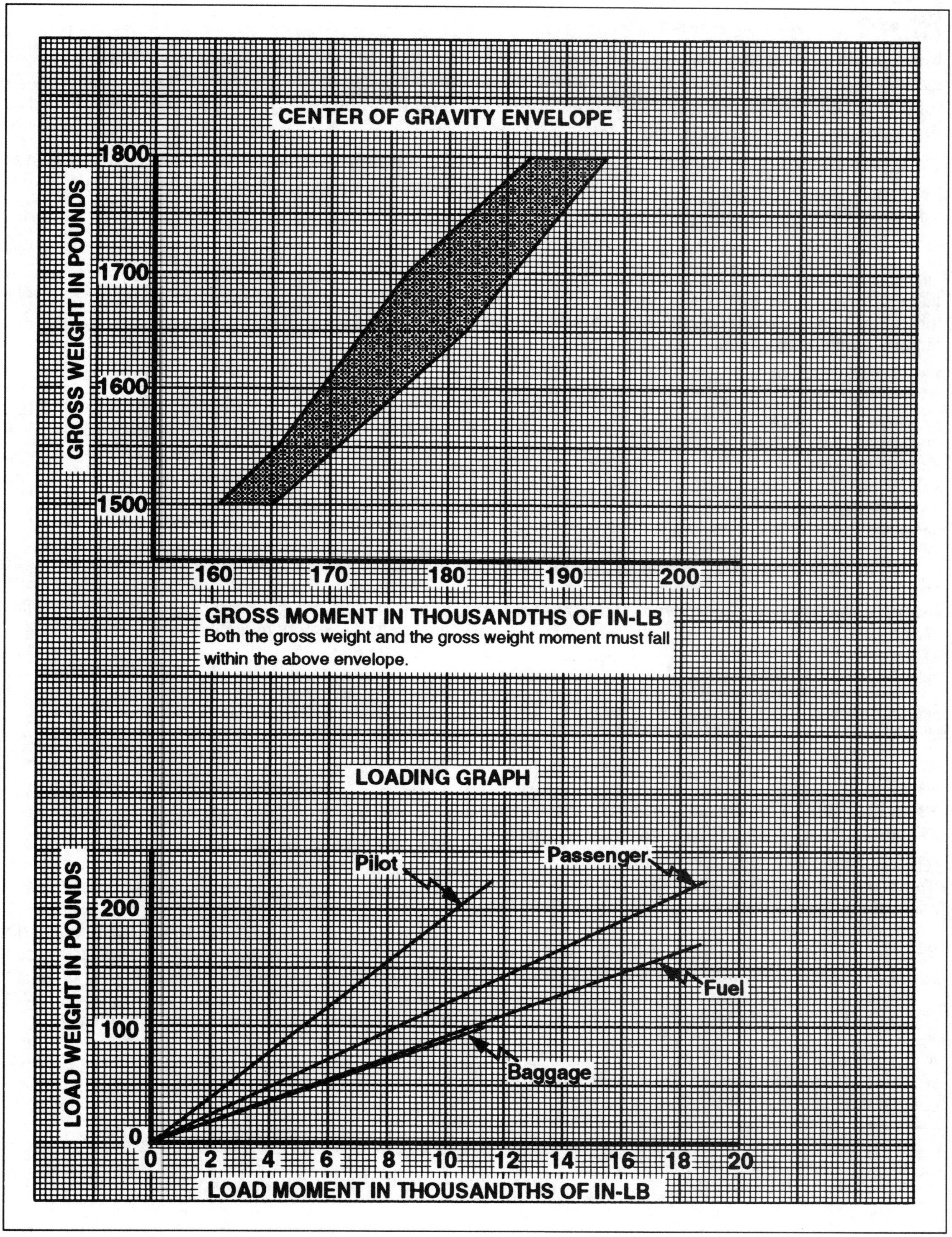

Center-of-Gravity Envelope and Loading Graph

Powerplant Systems

Most reciprocating engines used to power small aircraft incorporate two separate magneto ignition systems. A magneto ("mag") is a self-contained source of electrical energy, so even if an aircraft loses total electrical power, the engine will continue to run. The main advantages of the dual ignition system are increased safety and improved engine performance.

When checking for magneto operation prior to flight, the engine should run smoothly when operating the magneto selector set to "BOTH," and should experience a slight drop in revolutions per minute (RPM) when running on only one or the other magneto.

AIR, RTC, WSC, PPC
2238. An electrical system failure (battery and alternator) occurs in a magneto equipped aircraft during flight. In this situation, you would

A—probably experience engine failure due to the loss of the engine-driven fuel pump and also experience failure of the radio equipment, lights, and all instruments that require alternating current.
B—probably experience failure of the engine ignition system, fuel gauges, aircraft lighting system, and avionics equipment.
C—experience avionics equipment failure.

If you experience an inflight electrical system failure, you have an avionics equipment failure and cannot use your electrical fuel boost pump. (PLT207) — FAA-H-8083-25

Answer (A) is incorrect because the engine-driven fuel pumps are mechanical and not dependent on the electrical system. Answer (B) is incorrect because the ignition system of an aircraft reciprocating engine is powered by two self-contained magnetos that are not dependent upon the aircraft electrical system.

AIR, RTC, WSC, PPC
2333. An electrical system failure (battery and alternator) occurs during flight. In this situation, you would

A—experience avionics equipment failure.
B—probably experience failure of the engine ignition system, fuel gauges, aircraft lighting system, and avionics equipment.
C—probably experience engine failure due to the loss of the engine-driven fuel pump and also experience failure of the radio equipment, lights, and all instruments that require alternating current.

If you experience an infight electrical system failure, you would have an avionics equipment failure, and could not use the electrical fuel boost pump. (PLT207) — FAA-H-8083-25

Answer (B) is incorrect because the ignition system of an aircraft reciprocating engine is powered by two self-contained magnetos that are not dependent upon the aircraft electrical system. Answer (C) is incorrect because the engine-driven fuel pumps are mechanical and not dependent on the electrical system.

AIR, RTC, WSC, PPC
2343. One purpose of the dual ignition system on an aircraft engine is to provide for

A—improved engine performance.
B—uniform heat distribution.
C—balanced cylinder head pressure.

The dual ignition system has two magnetos to supply the electrical current to two spark plugs for each combustion chamber. This provides both a redundancy of ignition and an improvement of engine performance. (PLT478) — FAA-H-8083-25

Answer (B) is incorrect because heat distribution is not affected by the ignition system. Answer (C) is incorrect because cylinder head pressure in not affected by the ignition system.

AIR, RTC, WSC, PPC
2242. One purpose of the dual ignition system on a two-cycle engine is to provide for

A—system redundancy in the ignition system.
B—uniform heat distribution.
C—balanced cylinder head pressure.

The dual ignition system has two magnetos to supply the electrical current to two spark plugs for each combustion chamber. This provides both a redundancy of ignition and an improvement of engine performance. (PLT478) — FAA-H-8083-25

Answers
2238 [C] 2333 [A] 2343 [A] 2242 [A]

AIR, RTC, WSC, PPC

2243. The formation of ice in a carburetor's throat is indicated by

A—rough engine operation, followed by a decrease in oil pressure.
B—a rapid increase in RPM, followed by rough engine operation.
C—a drop in RPM, followed by rough engine operation.

Carb ice restricts the airflow into the engine reducing its power and resultant RPM, which also results in rough engine operation. (PLT190) — The Powered Parachute Bible, Chapter 5 and FAA-H-8083-25

Answers (A) and (B) are incorrect because oil pressure is not significantly affected by carb ice, and carb ice will not increase RPM.

AIR, RTC, WSC, PPC

2244. The purpose of the fuel tank vent system is to

A—remove dangerous vapors from the aircraft and prevent an explosion.
B—allow air to enter the tank as fuel is consumed.
C—ensure a proper fuel to air ratio.

Fuel tanks are not normally sealed systems; they need air venting because the fuel level is falling, therefore no vacuum builds up in the fuel tank. (PLT253) — The Powered Parachute Bible, Chapter 1

Answer (A) is incorrect because fuel tanks are isolated and fuel vapors are prevalent within the fuel tank confined area. Answer (C) relates to 2-stroke engines only. It is incorrect because fuel and oil are mixed either before the mixture is poured into the fuel tank (pre-mix), or after when oil is injected into the air/fuel mixture oil before it goes into the combustion chamber.

AIR, RTC, WSC, PPC

2245. A standby source of fuel to an engine in a powered parachute is typically

A—from an electrically powered pump.
B—through gravity feed.
C—from a pressurized fuel tank.

Some engines use an electric boost pump similar to an airplane to supply a back-up pump in case the engine-driven fuel pump fails when low to the ground. (PLT253) — The Powered Parachute Bible, Chapter 5

Answer (B) is incorrect because if a gravity feed system was used, then this would supply fuel and no standby pump would be needed. Answer (C) is incorrect because pressurized fuel tanks are not normally used for light-sport aircraft.

AIR, RTC, WSC, PPC

2246. The fuel vents on many powered parachutes and weight shift control aircraft are located

A—in the fuel cap.
B—adjacent to the crankcase breather.
C—in the fuel tank pressure relief valve.

The fuel vent in many but not all fuel tanks is in the fuel cap. (PLT253) — The Powered Parachute Bible, Chapter 1

Answer (B) is incorrect because the crankcase breather is used on a 4-stroke engine and has nothing to do with the fuel supply system on a 2-stroke engine. Answer (C) is incorrect because most PPCs do not have pressurized fuel tanks nor a fuel tank pressure relief valve.

AIR, RTC, WSC, PPC

2247. Combusted fuel is expelled from a 2-cycle engine through an

A—exhaust valve and exhaust port.
B—exhaust valve.
C—exhaust port.

All 2-stroke engines expel exhaust through a passage called the exhaust port. (PLT343) — The Powered Parachute Bible, Chapter 4

Answers (A) and (B) are incorrect because only some higher power engines use an exhaust valve, while all have an exhaust port.

AIR, RTC, WSC, PPC

2248. Fuel enters a two-cycle engine through an

A—intake port and intake valve.
B—intake port and reed valve.
C—intake valve and reed valve.

The air/fuel/oil mixture enters the crankcase through an intake port that some types of valve systems use to close off the crankcase and pressurize the air/fuel/oil mixture, before it is ported up to the top of the piston. Many engines use the positioning of the piston as the intake valve system, others use a rotary valve, while still others use a one-way flow "reed" or "poppet" valve. (PLT253) — The Powered Parachute Bible, Chapter 5

Answer (B) is incorrect because not all two-stroke engines use a reed valve. Answer (C) is incorrect because you must have an intake port.

Answers

2243 [C]	2244 [B]	2245 [A]	2246 [A]	2247 [C]	2248 [A]

AIR, RTC, WSC, PPC

2249. The first indication of carburetor ice in an aircraft with a four-cycle engine and fixed-pitch propeller is

A—an increase in RPM.
B—a decrease in RPM.
C—a decrease in oil pressure.

The first symptom of carb ice in a 4-stroke engine is a reduction in engine RPM. (PLT190) — The Powered Parachute Bible, Chapter 5

Answer (A) is incorrect because carb ice reduces RPM. Answer (C) is incorrect because carb ice has no noticeable effect on oil pressure.

AIR, RTC, WSC, PPC

2250. Air cooled engines dissipate heat

A—through cooling fins on the cylinder and head.
B—by air flowing through the radiator fins.
C—through the cylinder head temperature probe.

Air cooled engines use fins on the cylinder and head, forcing air past them as the primary means to dissipate heat. (PLT343) — The Powered Parachute Bible, Chapter 5

Answer (B) is incorrect because air cooled engines do not have separate radiators; only water and oil coolers need a radiator. Answer (C) is incorrect because the probe is not used to dissipate heat.

AIR, RTC, WSC, PPC

2251. Coolant in a liquid cooled engine is normally circulated by

A—capillary attraction.
B—an electric pump.
C—an engine driven pump.

Most liquid-cooled systems are driven from some mechanical source or pump on the engine. (PLT342) — The Powered Parachute Bible, Chapter 5

Answer (A) is incorrect because capillary attraction is not normally used for engine cooling. Answer (B) is incorrect because electric pumps usually are not used as coolant pumps.

AIR, RTC, WSC, PPC

2252. In order to improve engine efficiency, two-cycle engine exhaust systems are tuned to

A—close the exhaust valve to stop the fuel mixture from exiting the cylinder.
B—stop the fuel mixture from exiting the cylinder before combustion.
C—use a reed valve to stop the fuel mixture from exiting the cylinder.

If there is not an exhaust valve, tuned exhaust systems are designed to provide back pressure pulses at the exhaust port. The tuned exhaust bounces pressure back at the appropriate time, so the fuel mixture stays in the combustion chamber while both intake and exhaust ports are open. (PLT343) — The Powered Parachute Bible, Chapter 4

Answer (A) is incorrect because if the exhaust port has a valve, it is not as critical to have a tuned exhaust to provide back pressure at the exhaust port. There are not usually exhaust valves in a 2-stroke engine; however, some 2-stroke engines do have them. An example of this is the RAVE exhaust valve on the ROTAX 618. Answer (C) is incorrect because the reed valve is typically used for the intake air/fuel mixture.

AIR, RTC, WSC, PPC

2253. 2-cycle engine thrust and fuel efficiency can be greatly compromised when

A—exhaust systems are installed that are not specifically tuned for an engine.
B—carbon deposits build up on exhaust valves.
C—intake valve lifters fail to pressurize and provide adequate fuel to the combustion chamber.

The exhaust systems should be tuned to the engine for maximum efficiency. (PLT343) — The Powered Parachute Bible, Chapter 4

Answer (B) is incorrect because most 2-stroke engines do not have exhaust valves. Answer (C) is incorrect because not all 2-stroke engines have intake valve lifters.

Answers

2249 [B]	2250 [A]	2251 [C]	2252 [B]	2253 [A]

AIR, RTC, WSC, PPC

2254. The purpose of a kill switch is to

A—shut off the fuel to the carburetor.
B—ground the lead wire to the ignition coil shutting down the powerplant.
C—ground the battery eliminating current for the ignition system.

The kill switch shuts down the engine. (PLT207) — The Powered Parachute Bible, Chapter 6

Answer (A) is incorrect because a fuel valve shuts off fuel. Answer (C) is incorrect because the battery should already be grounded.

AIR, RTC, WSC, PPC

2255. A typical two-cycle engine ignition coil is powered by

A—a battery.
B—a battery or an alternator.
C—a magneto.

The magneto is an engine-driven generator that powers the ignition system and also supplies extra power to aircraft electrical system. (PLT478) — The Powered Parachute Bible, Chapter 7

Answers (A) and (B) are incorrect because the battery is further down the line, and is also powered by the magneto.

AIR, RTC, WSC, PPC

2256. Many 4-cycle engines utilize what type of lubrication system?

A—Forced.
B—Gravity.
C—Fuel/oil mixture.

Most 4-stroke engines have an oil pump that forces the oil through the system. (PLT324) — The Powered Parachute Bible, Chapter 5

Answer (B) is incorrect because gravity systems are not a typical oil supply system. Answer (C) is incorrect because a fuel/oil mixture system is on 2-stroke engines only.

AIR, RTC, WSC, PPC

2257. Adding more oil to the fuel than specified by the manufacturer of a 2-cycle engine will result in

A—increased engine performance.
B—increased carbon buildup and engine fouling.
C—increased engine lubrication and optimal performance.

Extra oil in the fuel would cause inefficient burning and more carbon buildup as a result. (PLT324) — The Powered Parachute Bible, Chapter 5

Answers (A) and (C) are incorrect because oil does not increase performance.

AIR, RTC, WSC, PPC

2258. Pilots should refrain from revving an engine with a reduction drive because

A—the crankshaft counterbalances may be dislodged and cause extreme engine vibration.
B—the propeller blade tips may exceed their RPM limits.
C—the torque exerted on the gears during excessive acceleration and deceleration can cause the gear box to self-destruct.

Revving the engine causes more stress than not revving it; it is good practice to not rev it unnecessarily. (PLT343) — The Powered Parachute Bible, Chapter 5

Answer (A) is incorrect because not all engines have counterbalances. Answer (B) is incorrect because the propeller is designed to not exceed its maximum RPM at full power.

AIR, RTC, WSC, PPC

2259. During preflight, the fuel vent system should always be checked

A—to ensure the vent is closed.
B—to ensure the vent is open.
C—to ensure the vent system pressure is in the green range.

The fuel vent needs to be open for flight for the air to fill the fuel tank as the fuel is consumed. (PLT253) — The Powered Parachute Bible, Chapter 1

Answer (A) is incorrect because a closed fuel vent would cause a power loss in flight when air is unable to fill the tank as fuel is used. Answer (C) is incorrect because there is no pressure in the fuel tank since it is vented.

Answers

2254 [B]	2255 [C]	2256 [A]	2257 [B]	2258 [C]	2259 [B]

AIR, RTC, WSC, PPC

2260. Carburetor ice can form

A—only at temperatures near freezing and the humidity near the saturation point.
B—when the outside air temperature is as high as 100 degrees F and the humidity is as low as 50%.
C—at any temperature or humidity level.

Carburetor ice can form when the outside air temperature is as high as 100°F and the humidity is as low as 50%. This is not the optimum conditions for the ice to form, but it can form under these conditions. (PLT190) — The Powered Parachute Bible, Chapter 5

Answer (A) is incorrect because carb ice can form when it is as high at 100°F. Answer (C) is incorrect because some moisture is needed to form the ice.

Category-Specific Operations

The explanations for the answers given describe the concepts that should be understood before taking the test. Many of these questions are based on older weight-shift and PPC designs and unique characteristics of specific designs. The answers given are the best of the choices provided.

Weight-Shift Control

WSC

2261. As a weight shift aircraft wing approaches a stall, the wing tips

A—decrease the wing's angle of attack.
B—act in much the same way as ailerons on a three-axis aircraft.
C—increase the wing's angle of attack.

As the angle of attack of the wing is increased, the nose is at a higher angle of attack and therefore stalls first while the tips keep flying. This drops the nose and as a result, decreases the wing's angle of attack. (PLT242) — Trikes, the Flex-Wing Flyers, Chapter 3

Answer (B) is incorrect because the wing tips act the same way ailerons do while flying normally and while approaching a stall. Answer (C) is incorrect because the tips only increase the wing's angle of attack when the wing is at a low angle of attack, far from a stall.

WSC

2262. During a wing stall, the wing tips of a weight shift aircraft are

A—ineffective for stall recovery.
B—effective for stall recovery.
C—effective only when combined with maximum engine output.

Since the wing tips are at a lower angle of attack, they do not normally stall when the rest of the wing is stalled. They keep flying, creating an up-force in back of the CG—causing the nose to rotate down and decrease the angle of attack of the wing—therefore they are very effective for stall recovery. (PLT242) — Trikes, the Flex-Wing Flyers, Chapter 3

Answer (A) is incorrect because the tips are very effective for stall recovery, allowing the nose to fall through. Answer (C) is incorrect because the tips have more effect for stall recovery than the engine power.

Answers

2260 [B]	2261 [A]	2262 [B]

WSC
2263. The crosstube is positioned by

A—a quick release pin.
B—self-locking bolts.
C—restraining cables attached to the rear of the keel.

The crossbar is pulled back to tension the airframe into the sail with the crossbar cables. These are attached to a connection point on the rear of the keel. (PLT114) — Trikes, the Flex-Wing Flyers, Chapter 3

Answers (A) and (B) are incorrect because these are fasteners and would only make smaller variations in the crossbar position if these fasteners were adjusted to different settings.

WSC
2347. One of the functions of the wing's crosstube is to

A—hold the wings open.
B—provide surface to grip and control the aircraft.
C—provide an attachment point for the carriage.

The purpose of the crosstube is to hold the leading edges open and therefore the wing. (PLT114) — Trikes, the Flex-Wing Flyers

Answer (B) is incorrect because it is the control bar you grip to control the aircraft. Answer (C) is incorrect because the attachment point for the carriage is the keel, the carriage cannot be attached to the crossbar.

WSC
2264. On some trikes, the hang point is part of

A—a variable trim arrangement that allows the pilot to adjust the aircraft center of gravity during flight to obtain the most favorable aircraft performance.
B—an adjustable trim arrangement that allows the pilot to adjust the aircraft center of gravity during flight to obtain the most favorable aircraft performance.
C—an adjustable trim arrangement that allows the center of gravity to shift fore and aft along the wing's keel.

Most trikes have an adjustment to move the position on the keel fore and aft on the ground. This is a common way to adjust the trim speed and bar position of the wing. (PLT114) — Trikes, the Flex-Wing Flyers, Chapter 3

Answers (A) and (B) are incorrect because although this is a viable design concept that has been and may be used for trikes, not many incorporate the complexity of a variable CG during flight.

WSC
2348. Which aircraft component ensures the wing has a pitch-up tendency?

A—Keel pocket.
B—Luff lines.
C—Washout rod.

The luff lines can be used for pitch up stability in level flight by pulling the trailing edge up, creating a reflex in the training edge of the wing from the root out towards the tips. The luff lines also provide a positive pitch up tendency in a steep dive or slight negative angles of attack by increasing reflex in the airfoil by keeping the trailing edge pulled up in this unusual situation. (PLT114) — Trikes, the Flex-Wing Flyers

Answer (A) is incorrect because the keel pocket simply holds the keel in place. Answer (C) is incorrect because the washout rods only provide a positive pitch-up moment when the aircraft is in a steep dive at zero angle of by holding the trailing edge up at the tips; the washout rods have no effect on the flight characteristics of the aircraft in normal flight conditions.

WSC
2265. The keel pocket's purpose is to

A—act as a longitudinal stabilizer, keeping the wing from wandering left and right.
B—act as a roll stabilizer, keeping the wing from wandering left and right.
C—act as a yaw stabilizer, keeping the wing from wandering left and right.

The most significant effect a keel pocket could have on stability would be for yaw. This was an early design concept used in the development of the flex wing. Today, the wing sweep, washout, and airfoil shape are designed to optimize the tracking (yaw) for the vertical axis. Keel pockets today are a fabric channel in the sail material the keel is inserted into, to hold the keel in place at the root of the wing. (PLT114) — Trikes, the Flex-Wing Flyers, Chapter 3

Answers (A) and (B) are incorrect because the keel pocket does not supply this stability.

Answers
2263 [C] 2347 [A] 2264 [C] 2348 [B] 2265 [C]

WSC

2266. How does the wing design feature "washout" affect the production of lift?

A—The wing tips continue producing lift when the main body of the wing is not producing lift.
B—The main body of the wing continues to produce lift when the wing tips are not producing lift.
C—The center of lift moves from the trailing edge of the wing, to the leading edge of the wing, as the wing begins to stall.

The washout/twist in the wing, starts with a high angle of attack at the root/nose, and decreases the angle of attack as the you approach each tip. This washout/twist, sweep, and airfoil shape is designed into the wing to make the nose lose lift first while the tips keep flying at high angles of attack. (PLT114) — Trikes, the Flex-Wing Flyers, Chapter 3

Answer (B) is incorrect because this happens only when the wing is at very low angles of attack where the wing is not near the critical angle of attack to stall. Answer (C) is incorrect because this is not the design of any trike wing and would produce a wing that would be unstable near the stall.

WSC

2267. The wing of a weight-shift aircraft twists so that the angle of attack

A—from the center of the wing to the wing tip is variable and can be adjusted by the pilot in flight to optimize performance.
B—changes from a low angle of attack at the center of the wing, to a high angle of attack at the tips.
C—changes from a high angle of attack at the center of the wing, to a low angle of attack at the tips.

The fundamental design of the flex wing is for the wing to twist from a high angle of attack at the nose, to a lower angle of attack at the tips. (PLT214) — Trikes, the Flex-Wing Flyers, Chapter 3

Answer (A) is incorrect because this would provide a wing that would be unstable and dangerous. Flex wings are not designed this way. Answer (B) is incorrect because this only applies for some wing designs but not all. Varying the twist in the wing is common to most high performance hang gliders in flight and used as one method to trim weight-shift wings as well.

Powered Parachute

PPC

2268. During flight, advancing thrust will

A—increase airspeed.
B—cause the aircraft to climb.
C—cause the aircraft to increase airspeed and climb.

Throttle controls vertical speed in a PPC. Advancing the throttle will produce decreased descent rates or increased climb rates. Speed in a PPC is controlled by the weight and not the throttle. (PLT114) — The Powered Parachute Bible, Chapter 1

Answers (A) and (C) are incorrect because throttle does not affect airspeed.

PPC

2269. The torque effect of an engine that rotates clockwise in a powered parachute is counteracted by

A—increasing the length of the right and decreasing the length of the left riser cables.
B—decreasing the length of the left riser cables.
C—decreasing the length of right riser cables.

A clockwise or right-turning propeller when viewed from the rear creates an opposite reaction to turn the undercarriage aircraft to the left. Therefore, a slight right-hand turn needs to be built into the aircraft to accommodate for this torque. Many designs are used by manufacturers to accomplish this. Decreasing the length of the right-hand riser will accomplish this by bringing the right side of the wing down. (PLT114) — The Powered Parachute Bible, Chapter 2

Answers (A) and (B) are incorrect because they would create a turn in the wrong direction.

PPC

2270. The steering bars

A—are used during taxi operations with the parachute stowed.
B—control the outboard trailing edge of the parachute.
C—control the main landing gear brakes.

The steering bars are the main control to turning in flight. Pushing on the right-hand steering bar will pull the right control line, lower the trailing edge of the right wing, create more drag on the right side and turn the aircraft to the right. (PLT346) — The Powered Parachute Bible, Chapter 1

Answers (A) and (C) are incorrect because the steering bars control the wing and are not used for ground operations.

Answers

2266 [A]	2267 [C]	2268 [B]	2269 [C]	2270 [B]

PPC
2271. The center of gravity tube is

A—lengthened for heavier pilots.
B—shortened for lighter pilots.
C—lengthened for lighter pilots.

The lighter the pilot, the more rearward the wing attachment should be for the hanging airframe to be balanced properly. Most modern designs have a number of hook-in points fore and aft on the airframe. For the tube CG adjustment system, lengthening the CG tube moves the wing hang point back to account for the lightweight person in the front. (PLT114) — The Powered Parachute Bible, Chapter 3

Answer (A) is incorrect because this would balance the airframe with the nose wheel too low. Answer (B) is incorrect because this would move the CG forward and the front wheel would be too high.

PPC
2272. The fan guard surrounds the propeller and

A—increases aerodynamic efficiency.
B—reduces "P" factor.
C—protects the parachute suspension lines from damage.

The purpose of the fan guard is to protect the parachute lines from hitting the prop. (PLT114) — The Powered Parachute Bible, Chapter 1

Answers (A) and (B) are incorrect because the fan guard reduces performance and has no effect on P factor.

PPC
2273. Cross ports in the parachute ribs aid in

A—weight reduction of the canopy.
B—the pressurization of the neighboring cells.
C—drying of the canopy.

Cross ports in the wing ribs allow air to flow sideways from cell to cell, called "cross flow" in the wing. This causes the cells next to each other to transfer pressure inside the wing and cells to pressurize neighboring cells. (PLT114) — The Powered Parachute Bible, Chapter 2

PPC
2274. Splicing severed suspension lines

A—is permissible if using the same size material as the original line.
B—is a very dangerous practice.
C—is an acceptable field repair.

Splicing lines is dangerous because you can change the airfoil; the lines could come loose and go through the prop. (PLT114) — The Powered Parachute Bible, Chapter 2

Answers (A) and (C) are incorrect because it is a dangerous practice.

PPC
2275. Tying a severed suspension line

A—will change the shape of the wing and is not permissible.
B—is permissible if it is shortened no more than six inches.
C—is an acceptable field repair.

Tying a severed suspension line would shorten it and create a discontinuity in the wing shape and is not acceptable. (PLT114) — The Powered Parachute Bible, Chapter 2

Answers (B) and (C) are incorrect because it is a dangerous practice.

PPC
2276. What gives your powered parachute wing/canopy its airfoil shape?

A—The risers because, by decreasing the length of the right riser you will get the precise airfoil shape.
B—The suspension lines as they are precisely measured and fitted to a specific location.
C—The air as it enters the cell openings on the leading edge of the airfoil.

The precise lengths of the suspension lines determine the bottom shape of the airfoil. Airfoil-shaped ribs attached to the bottom of the airfoil where the suspension lines are attached define the top shape of the airfoil. (PLT114) — The Powered Parachute Bible

Answer (A) is incorrect because the decreased length of the right riser is used to counteract torque on some designs, thus causing a turn. Answer (C) is incorrect because the air inflates the canopy to whatever airfoil shape the suspension lines and ribs define.

Answers

2271 [C]	2272 [C]	2273 [B]	2274 [B]	2275 [A]	2276 [B]

PPC
2277. Swapping wings from one brand or type of powered parachute to another is

A—permissible as long as the basic shape of the parachutes are similar.
B—dangerous since every wing is designed for a specific aircraft.
C—permissible if the overall area of the parachutes is the same.

Every wing is designed for a specific airframe configuration, engine torque and weight. (PLT114) — The Powered Parachute Bible, Chapter 2

Answers (A) and (C) are incorrect because there can be a different length in the lines between the right and left, which has nothing to do with the shape and area of the wing.

PPC
2278. Degradation of the parachute's protective polyurethane coating results in

A—increased takeoff distances, decreased maximum gross weight, and increased fuel consumption.
B—reduced takeoff distances, increased maximum gross weight, and reduced fuel consumption.
C—increased takeoff distances, increased maximum gross weight, and increased fuel consumption.

A degradation of the fabric results in air leaking through the fabric and a loss in performance since this creates more drag. (PLT114) — The Powered Parachute Bible, Chapter 2

Answers (B) and (C) are incorrect because there is no increased gross weight.

PPC
2279. Flaring allows the pilot to touchdown at a

A—higher rate of speed and a slower rate of descent.
B—lower rate of speed and a higher rate of descent.
C—lower rate of speed and a lower rate of descent.

Flaring slows you down and decreases your descent rate for a soft and slow landing. (PLT221) — The Powered Parachute Bible, Chapter 1

Answers (A) and (B) are incorrect because flaring does not produce a higher rate of speed or a higher rate of descent.

PPC
2280. Flaring during a landing

A—decreases the powered parachute's speed due to increased drag.
B—increases the powered parachute's speed due to reduced drag.
C—decreases the powered parachute's drag due to increased speed.

Flaring or pulling down on the trailing edge creates more drag and slows the aircraft similar to a flap on an airplane. (PLT221) — The Powered Parachute Bible, Chapter 1

Answers (B) and (C) are incorrect because flaring does not increase the speed of a PPC.

Answers

2277 [B]	2278 [A]	2279 [C]	2280 [A]

Balloon

LTA

2284. The term "weigh-off" means to determine the

A—static equilibrium of the balloon as loaded for flight.
B—amount of gas required for an ascent to a preselected altitude.
C—standard weight and balance of the balloon.

The term "weigh-off" means to determine the neutral buoyancy of a gas balloon or airship by taking weight off at launch. "Weigh-off" is used to determine the static equilibrium of the balloon. (PLT267) — FAA-H-8083-11, Glossary

LTA

2285. The lifting forces which act on a hot air balloon are primarily the result of the interior air temperature being

A—greater than ambient temperature.
B—less than ambient temperature.
C—equal to ambient temperature.

A hot air balloon derives lift from the fact that air inside the envelope is warmer and therefore "lighter" than the air around the balloon. (PLT181) — FAA-H-8083-11

LTA

2287. On cold days, it may be necessary to preheat the propane tanks because

A—the temperature of the liquid propane controls the burner pressure during combustion.
B—there may be ice in the lines to the burner.
C—the propane needs to be thawed from a solid to a liquid state.

On very cold days, it may be necessary to preheat the propane tanks since the temperature of the liquid propane controls the burner pressure during combustion. (PLT254) — FAA-H-8083-11

LTA

2288. How should a balloon fuel system be checked for leaks prior to flight?

A—Listen and smell.
B—Check all connections with a lighted match.
C—Cover all connections and tubing with soapy water.

Propane is under pressure and has an artificial odor. If a leak exists, the fuel will "hiss" out and can be smelled. (PLT251) — Balloon Ground School

LTA

2289. In addition to the required documents, what carry-on equipment should be accounted for during preflight?

A—Flotation gear.
B—Emergency locator transmitter.
C—Two means of burner ignition.

As a precaution against flameout, an ignitor must be carried to relight the burner. A second ignitor is necessary in case the first malfunctions or is lost overboard. (PLT253) — FAA-H-8083-11

LTA

2290. All fuel tanks should be fired during preflight to determine

A—the burner pressure and condition of the valves.
B—that the pilot light functions properly on each tank.
C—if there are any leaks in the tank.

Burner output is dependent on fuel pressure. Proper valve operation is critical to safety and control. (PLT254) — FAA-H-8083-11

LTA

2291. If ample propane is available, within which temperature range will propane vaporize sufficiently to provide enough pressure for burner operation during flight?

A—0 to 30°F.
B—10 to 30°F.
C—30 to 90°F.

When ample liquid propane is available, propane will vaporize sufficiently to provide proper operation between 30°F and 90°F. (PLT251) — Balloon Ground School

Answers

2284 [A] 2285 [A] 2287 [A] 2288 [A] 2289 [C] 2290 [A]
2291 [C]

LTA

2292. While in flight, ice begins forming on the outside of the fuel tank in use. This would most likely be caused by

A—water in the fuel.
B—a leak in the fuel line.
C—vaporized fuel instead of liquid fuel being drawn from the tank into the main burner.

If large quantities of vapor are withdrawn rapidly from a propane cylinder, the cooling effect of the vaporization in the tank cools the propane and lowers the vapor pressure and the rate of vaporization. (PLT254) — Balloon Ground School

LTA

2293. Burner efficiency of a hot air balloon decreases approximately what percent for each 1,000 feet above MSL?

A—4 percent.
B—8 percent.
C—15 percent.

Burner efficiency of a hot air balloon system decreases at approximately 4% per 1,000 feet above MSL. (PLT130) — Balloon Ground School

LTA

2295. On a balloon equipped with a blast valve, the blast valve is used for

A—climbs and descents only.
B—altitude control.
C—emergencies only.

The blast valve is located on the burner and controls ascent or descent by use of short bursts of power. When reaching the desired altitude, the pilot should reduce the frequency of blasts. As the envelope cools, the desired altitude can be maintained by using short blasts of heat evenly spaced. (PLT177) — FAA-H-8083-11

LTA

2296. When landing a free balloon, what should the occupants do to minimize landing shock?

A—Be seated on the floor of the basket.
B—Stand with knees slightly bent, in the center of the gondola, facing the direction of movement.
C—Stand back-to-back and hold onto the load ring.

By facing forward with knees bent, the body is balanced and the legs act as springs, absorbing the landing shock. (PLT184) — How to Fly a Balloon

LTA

2297. What causes false lift which sometimes occurs during launch procedures?

A—Closing the maneuvering vent too rapidly.
B—Excessive temperature within the envelope.
C—Venturi effect of the wind on the envelope.

False lift is caused by the venturi effect produced by the wind blowing across an inflated but stationary envelope. This is "dynamic lift" created by relative air movement. If the balloon is released, the relative wind decreases as the balloon accelerates to the speed of the wind and false lift decreases. (PLT304) — Powerline Excerpts

LTA

2298. What is a potential hazard when climbing at maximum rate?

A—The envelope may collapse.
B—Deflation ports may be forced open.
C—The rapid flow of air may extinguish the burner and pilot light.

The positive pressure on the top of the envelope could force the deflation ports open. (PLT183) — FAA-H-8083-11

Answers

2292 [C]	2293 [A]	2295 [B]	2296 [B]	2297 [C]	2298 [B]

LTA

2299. It may be possible to make changes in the direction of flight in a hot air balloon by

A—flying a constant atmospheric pressure gradient.
B—operating at different flight altitudes.
C—operating above the friction level, if there is no gradient wind.

A free balloon has no propulsion and so must take advantage of differing wind directions at various altitudes. (PLT219) — FAA-H-8083-11

LTA

2300. What is a hazard of rapid descents?

A—Wind shear can cavitate one side of the envelope, forcing air out of the mouth.
B—The pilot light cannot remain lit with the turbulent air over the basket.
C—Aerodynamic forces may collapse the envelope.

Turbulent airflow may extinguish the pilot light. (PLT125) — FAA-H-8083-11

LTA

2301. In a balloon, best fuel economy in level flight can be accomplished by

A—riding the haze line in a temperature inversion.
B—short blasts of heat at high frequency.
C—long blasts of heat at low frequency.

The desired envelope temperature for level flight is best maintained by short blasts at high frequency. (PLT130) — FAA-H-8083-11

LTA

2310. What constitutes the payload of a balloon?

A—Total gross weight.
B—Total weight of passengers, cargo, and fuel.
C—Weight of the aircraft and equipment.

Payload is the total weight of passengers, cargo, and fuel an aircraft can legally carry on a given flight, excluding weight of the aircraft and equipment. (PLT179) — Balloon Ground School

LTA

2313. Under which condition will an airship float in the air?

A—When buoyant force equals horizontal equilibrium existing between propeller thrust and airship drag.
B—When buoyant force is less than the difference between airship weight and the weight of the air volume being displaced.
C—When buoyant force equals the difference between airship weight and the weight of the air volume being displaced.

A lighter-than-air craft is in equilibrium when buoyancy equals weight. The buoyant force is equal to the weight of the air volume displaced. (PLT239) — Goodyear Airship Operations Manual

LTA

2314. The pressure height of an airship is the altitude at which

A—the airship would be unable to gain more altitude.
B—gas pressure would reach 3 inches of water.
C—the ballonet(s) would be empty.

Expansion of the gas will cause the ballonets to completely deflate at pressure height. Thus, no further increase in displaced air volume is possible. (PLT159) — Goodyear Airship Operations Manual

LTA

2315. If an airship should experience failure of both engines during flight and neither engine can be restarted, what initial immediate action must the pilot take?

A—The airship must be driven down to a landing before control and envelope shape are lost.
B—The emergency auxiliary power unit must be started for electrical power to the airscoop blowers so that ballonet inflation can be maintained.
C—Immediate preparations to operate the airship as a free balloon are necessary.

An airship without power must be operated as a free balloon. (PLT208) — Goodyear Airship Operations Manual

Answers

2299 [B] 2300 [B] 2301 [B] 2310 [B] 2313 [C] 2314 [C]
2315 [C]

LTA

2317. During flight in an airship, when is vertical equilibrium established?

A—When buoyancy is greater than airship weight.
B—When buoyancy equals airship weight.
C—When buoyancy is less than airship weight.

A lighter-than-air craft is in equilibrium when buoyancy equals weight. The buoyant force is equal to the weight of the air volume displaced. (PLT153) — Goodyear Airship Operations Manual

LTA

2322. An unbalanced condition of an airship in flight must be overcome by

A—valving air from the ballonets.
B—valving gas from the envelope.
C—a negative or a positive dynamic force.

Dynamic force, created by movement through the air, must be used to overcome any out-of-equilibrium condition. (PLT160) — Goodyear Airship Operations Manual

Gyroplane

RTC

2326. Why should gyroplane operations within the cross-hatched portion of a Height vs. Velocity chart be avoided?

A—The rotor RPM may build excessively high if it is necessary to flare at such low altitudes.
B—Sufficient airspeed may not be available to ensure a safe landing in case of an engine failure.
C—Turbulence near the surface can dephase the blade dampers causing geometric unbalanced conditions on the rotor system.

The chart can be used to determine those altitude-airspeed combinations from which it would be impossible to successfully complete an autorotative landing. The altitude-airspeed combinations that should be avoided are represented by the shaded areas of the chart. (PLT123) — FAA-H-8083-21

RTC

2327. The principal factor limiting the never-exceed speed (V_{NE}) of a gyroplane is

A—turbulence and altitude.
B—blade-tip speed, which must remain below the speed of sound.
C—lack of sufficient cyclic stick control to compensate for dissymmetry of lift or retreating blade stall, depending on which occurs first.

Retreating blade stall is the principal factor limiting the never-exceed speed. (PLT260) — FAA-H-8083-21

RTC

2328. Select the true statement concerning gyroplane taxi procedures.

A—Taxi speed should be limited to no faster than a brisk walk in ideal conditions.
B—The cyclic stick should be held in the neutral position at all times.
C—The cyclic stick should be held slightly aft of neutral at all times.

A gyroplane should not be taxied in close proximity to people or obstructions while the rotor is turning. In addition, taxi speed should be limited to no faster than a brisk walk in ideal conditions, and adjusted appropriately according to the circumstances. (PLT149) — FAA-H-8083-21

RTC

2329. If ground resonance is experienced during rotor spin-up, what action should you take?

A—Taxi to a smooth area.
B—Make a normal takeoff immediately.
C—Close the throttle and slowly raise the spin-up lever.

A corrective action for ground resonance is an immediate takeoff if RPM is in proper range (for helicopters) or an immediate closing of the throttle and placing the blades in low pitch if the RPM is low. "During spin-up" implies low RPM, so closing the throttle is appropriate. (PLT259) — FAA-H-8083-21

Answers

2317 [B]	2322 [C]	2326 [B]	2327 [C]	2328 [A]	2329 [C]

RTC

2330. If the pilot experiences ground resonance, and the rotor RPM is not sufficient for flight,

A—open the throttle full and liftoff.
B—apply the rotor brake and stop the rotor as soon as possible.
C—attempt to takeoff at that power setting.

Ground resonance is an aerodynamic phenomenon associated with fully-articulated rotor systems. It develops when the rotor blades move out of phase with each other and cause the rotor disc to become unbalanced. This condition can cause a helicopter to self-destruct in a matter of seconds. However, for this condition to occur, the helicopter must be in contact with the ground. If you experience ground resonance, and the rotor RPM is not yet sufficient for flight, apply the rotor brake to maximum and stop the rotor as soon as possible. If ground resonance occurs during takeoff, when RPM is sufficient for flight, lift off immediately. (PLT259) — FAA-H-8083-21

RTC

2331. Select the true statement concerning gyroplane taxi procedures.

A—Avoid abrupt control movements when blades are turning.
B—The cyclic stick should be held in the neutral position at all times.
C—The cyclic stick should be held slightly aft of neutral at all times.

A gyroplane should not be taxied in close proximity to people or obstructions while the rotor is turning. In addition, taxi speed should be limited to no faster than a brisk walk in ideal conditions, and adjusted appropriately according to the circumstances. Avoid abrupt control motions while taxiing. (PLT149) — FAA-H-8083-21

RTC

2332. During the transition from pre-rotation to flight, all rotor blades change pitch

A—simultaneously to the same angle of incidence.
B—simultaneously but to different angles of incidence.
C—to the same degree at the same point in the cycle of rotation.

Compensation for dissymmetry of lift requires constant change in the blade angle of incidence, with one increasing as another simultaneously decreases. During the transition from prerotation to flight (or any time there is dissymmetry of lift) all rotor blades change pitch simultaneously, but to different angles of incidence. (PLT470) — FAA-H-8083-21

RTC

2349. What precaution should be taken while taxiing a gyroplane?

A—The cyclic stick should be held in the neutral position at all times
B—Avoid abrupt control movements when blades are turning.
C—The cyclic stick should be held slightly aft of neutral at all times.

Avoid abrupt control motions while taxiing. (PLT149) — FAA-H-8083-21

Answers

2330 [B] 2331 [A] 2332 [B] 2349 [B]

RTC

2361. (Refer to Gyroplane Takeoff and Landing Graphs Figure below.) Determine the total takeoff distance required for a gyroplane to clear a 50-foot obstacle if the temperature is 95°F and the pressure altitude is 1,700 feet.

A—1,825 feet.
B—1,910 feet.
C—2,030 feet.

Use the following steps:

1. *Locate the 1,700-foot pressure altitude position on the Takeoff Distance Graph (at right).*
2. *Draw a horizontal line from the 1,700-foot position to the 95°F OAT point.*
3. *Draw a vertical line downward from the point of intersection to the total Takeoff Distance Scale.*

Note the distance is 2,030 feet.

(PLT011) — FAA-H-8083-21

RTC

2362. (Refer to Gyroplane Takeoff and Landing Graphs Figure below.) Determine the total landing distance to clear a 50-foot obstacle in a gyroplane. The outside air temperature (OAT) is 75°F and the pressure altitude at the airport is 2,500 feet.

A—521 feet.
B—525 feet.
C—529 feet.

Use the following steps:

1. *Locate the horizontal line corresponding to 2,500 feet pressure altitude on Gyroplane Takeoff and Landing Graphs figure.*
2. *Draw vertical lines through the points of intersection of the 2,500-foot pressure altitude line with the 75°F point. Note the landing distance of 525 feet.*

(PLT011) — FAA-H-8083-21

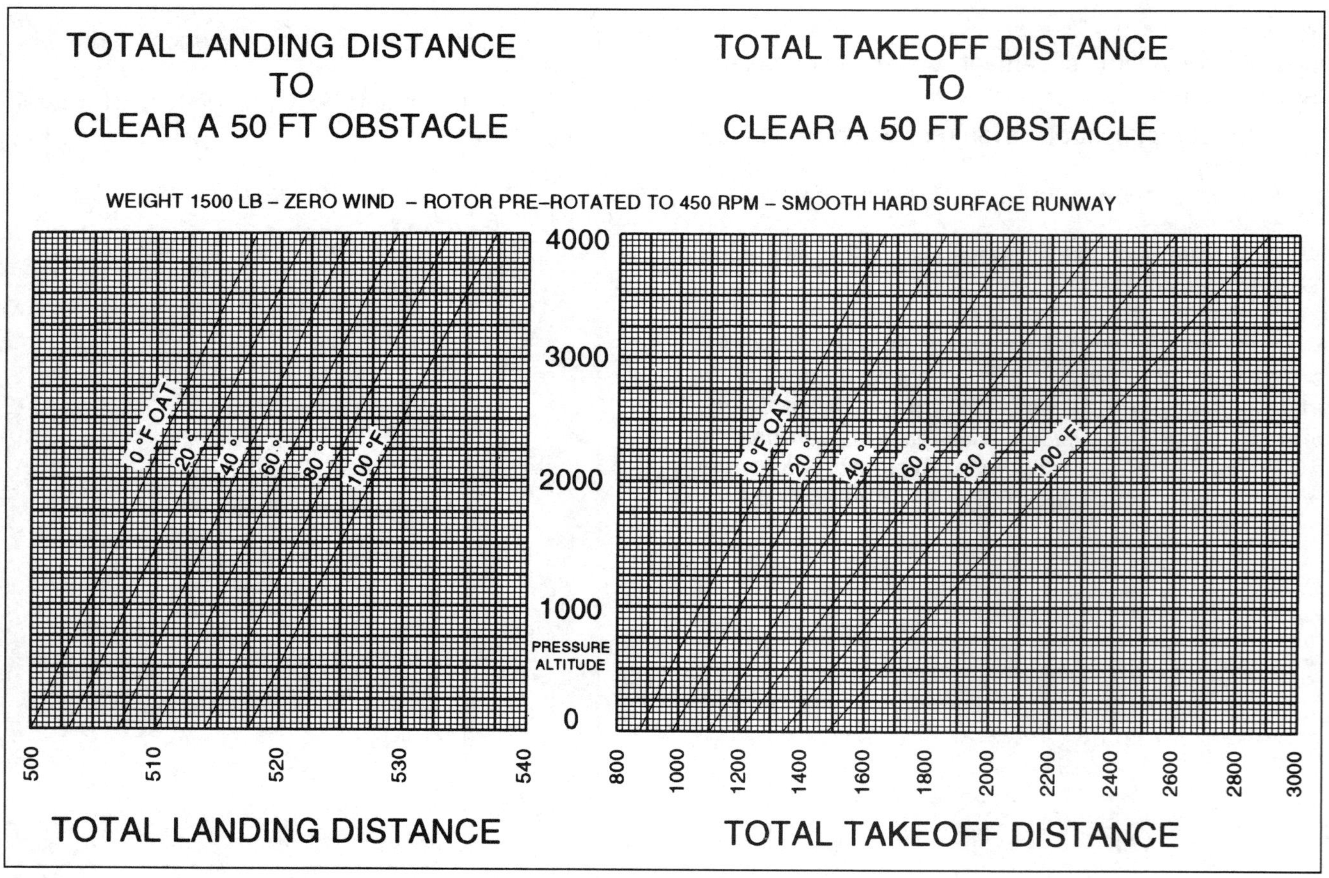

Gyroplane Takeoff and Landing Graphs

Answers
2361 [C] 2362 [B]

GLI

2359. (Refer to Glider Performance Graph Figure below.) What approximate lift/drag ratio will the glider attain at 68 MPH in still air?

A—10.5:1.
B—21.7:1.
C—28.5:1.

Use the following steps:

1. *Locate 68 MPH on the V coordinate (at the bottom) on Glider Performance Graph figure.*
2. *Read upward from the 68 MPH point to the intersection with the L/D curve.*
3. *Read to the left, from the point of intersection, to the L/D scale and read L/D, 28.5:1.*

(PLT017) — FAA-H-8083-13

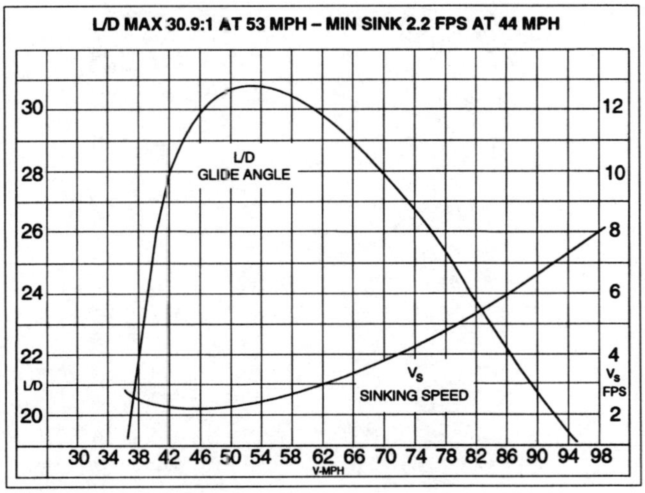

Glider Performance Graph

GLI

2360. What corrective action should the sailplane pilot take during takeoff if the towplane is still on the ground and the sailplane is airborne and drifting to the left?

A—Crab into the wind by holding upwind (right) rudder pressure.
B—Crab into the wind so as to maintain a position directly behind the towplane.
C—Establish a right wing low drift correction to remain in the flightpath of the towplane.

Crab into the wind to maintain a position directly behind the towplane. (PLT222) — FAA-H-8083-13

Answers
2359 [C] 2360 [B]

Chapter 5
Airports, Airspace and Charts

Uncontrolled and Tower-Controlled Airports

Airport Traffic Control Towers (ATCT) are established to promote the safe, orderly, and expeditious flow of air traffic. The tower controller will issue instructions for aircraft to follow the desired flight path while in the airport traffic area whenever necessary by using the following terminology: upwind leg, crosswind leg, downwind leg, base leg. *See* Figure 5-1.

The ATCT will also direct aircraft taxiing on the surface movement area of the airport. In all instances, an appropriate clearance must be received from the tower before taking off or landing.

At airports without an operating control tower, pilots of fixed-wing and weight-shift control aircraft must circle the airport to the left ("left traffic") unless sectionals, Airport/Facility Directories, or visual indicators indicate right traffic.

A common visual indicator is the segmented circle system, which consists of the following components:

- The **segmented circle** is located in a position readily visible to pilots in the air and on the ground. *See* Figure 5-2.
- A **tetrahedron** may be used to indicate the direction of landings and takeoffs. The small end of the tetrahedron points in the direction of landing. Pilots are cautioned against using the tetrahedron to determine wind direction, because it may not indicate the correct direction in light winds.
- A **wind cone, wind sock**, or **wind tee** may be installed near the operational runway to indicate wind direction. The large end of the wind cone or sock points into the wind as does the cross bar of the wind tee. The tetrahedron, wind cone, wind sock, or wind tee may be located in the center of the segmented circle and may be lit for night operations. *See* Figure 5-3.

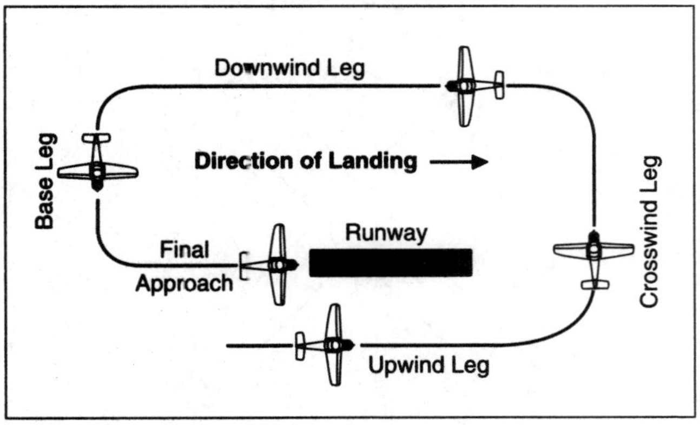

Figure 5-1

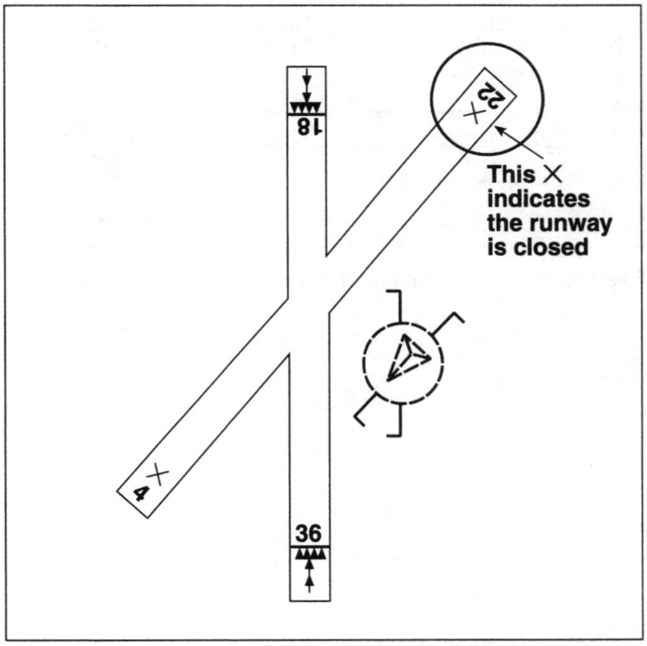

Figure 5-2. Segmented circle and landing direction indicator

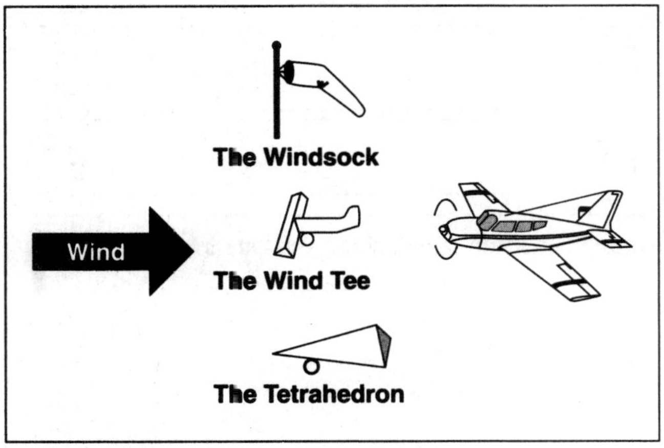

Figure 5-3. Wind/landing direction indicators

The designated beginning of the runway that is available and suitable for the landing of aircraft is called the **threshold** (Figure 5-4a). A threshold that is not at the beginning of the full-strength runway pavement is a **displaced threshold**. The paved area behind the displaced threshold is marked by arrows (Figure 5-4b) and is available for taxiing, takeoff, and landing rollout, but is not to be used for landing, usually because of an obstruction in the approach path. *See* Figure 5-4.

Stopways are found extending beyond some usable runways. These areas are marked by chevrons, and while they appear usable, they are suitable only as overrun areas. *See* Figure 5-5.

Runway numbers and letters are determined from the approach direction. The number is the magnetic heading of the runway rounded to the nearest 10°. For example, an azimuth of 183° would result in a runway number of 18; a magnetic azimuth of 076° would result in a runway numbered 8. Runway letters differentiate between left (L), right (R), or center (C). *See* Figure 5-6.

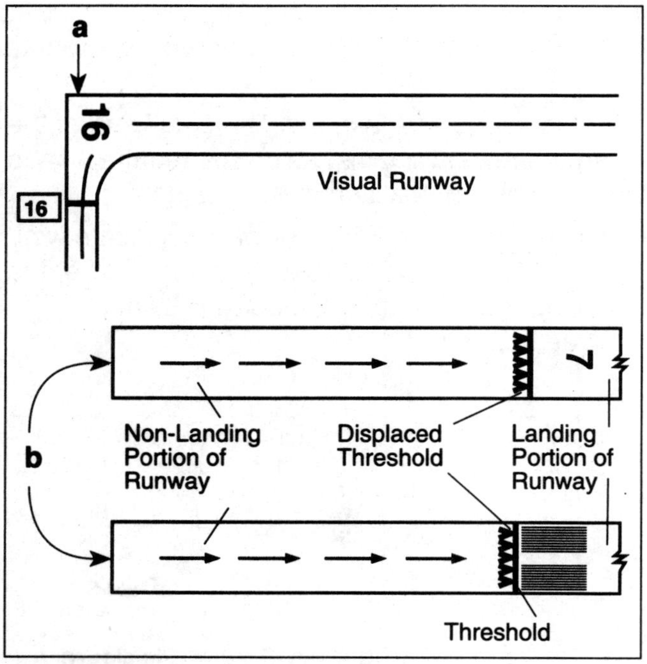

Figure 5-4. Threshold marking

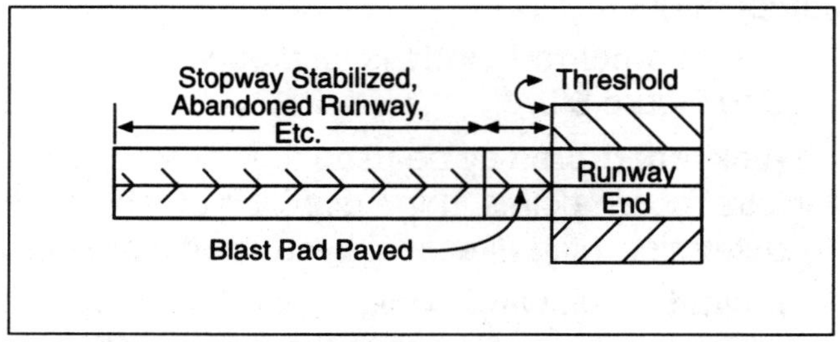

Figure 5-5. Stopway marking

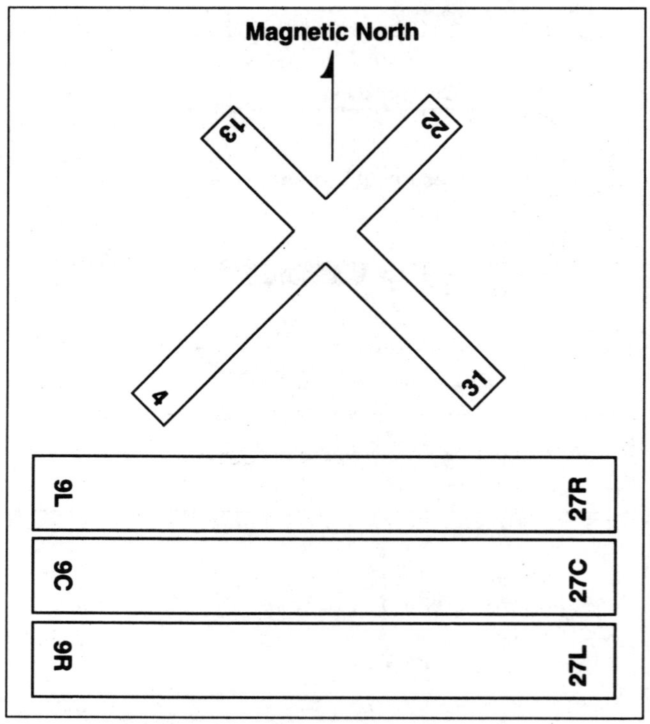

Figure 5-6. Runway numbers and letters

Landing runway (landing strip) indicators are installed in pairs and used to show alignment of runways. *See* Figure 5-7a. Traffic pattern indicators are installed in pairs in conjunction with landing strip indicators, and are used to indicate the direction of turns. *See* Figure 5-7b.

Approaching to land at an airport without a control tower, or when the control tower is not in operation, the pilot should observe the indicator for the approach end of the runway to be used.

Aircraft departing an uncontrolled airport must comply with any FAA traffic pattern established for that airport.

Entries into traffic patterns while descending create specific collision hazards and should be avoided.

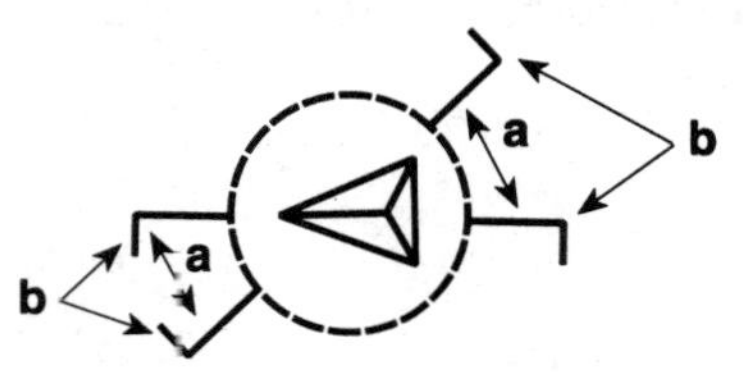

Figure 5-7. Landing runway and traffic pattern indicators

ALL

2035. Which is the correct traffic pattern departure procedure to use at a non-towered airport?

A—Depart in any direction consistent with safety, after crossing the airport boundary.
B—Make all turns to the left.
C—Comply with any FAA traffic pattern established for the airport.

In the case of an aircraft departing an airport without an operating control tower, comply with any FAA traffic pattern for that airport. (PLT201) — 14 CFR §91.127

ALL

2036. (Refer to Figure 63.) If the wind is as shown by the landing direction indicator, the pilot should land on

A—Runway 18 and expect a crosswind from the right.
B—Runway 22 directly into the wind.
C—Runway 36 and expect a crosswind from the right.

The small end of the tetrahedron points into the wind indicating the direction of landing. Landing to the south on RWY 18, the pilot could expect a right crosswind. (PLT039) — AIM ¶4-3-4

Answer (B) is incorrect because Rwy 22 is closed, as indicated by the X. Answer (C) is incorrect because the wind is from the southwest and you should land into the wind.

ALL

2037. (Refer to Figure 64.) Which runway and traffic pattern should be used as indicated by the wind cone in the segmented circle?

A—Right-hand traffic on Runway 9.
B—Right-hand traffic on Runway 18.
C—Left-hand traffic on Runway 36.

The large end of the wind cone (wind sock) points into the wind. The wind cone in FAA Figure 64 indicates a wind from the northwest. Landing into the wind can be accomplished on either Runway 27 or Runway 36. The traffic pattern indicators require right traffic to Runway 27 and left traffic to Runway 36. (PLT039) — AIM ¶4-3-4

ALL

2060. When approaching to land at an airport in Class G airspace that does not have light signals or other visual markings, an airplane pilot must make

A—a straight-in approach.
B—all turns to the right.
C—all turns to the left.

When approaching to land at an airport without an operating control tower in a Class G airspace area, each pilot of an airplane must make all turns of that airplane to the left unless the airport displays approved light signals or visual markings indicate otherwise. (PLT435) — 14 CFR §91.126

Answers

2035 [C]	2036 [A]	2037 [C]	2060 [C]

ALL
2194. When approaching to land at an airport, without an operating control tower, in Class G airspace, the pilot should

A—enter and fly a traffic pattern at 800 feet AGL.
B—make all turns to the left, unless otherwise indicated.
C—fly a left-hand traffic pattern at 800 feet AGL.

When approaching to land at an airport without an operating control tower in Class G airspace, each pilot of an airplane must make all turns to the left unless the airport displays approved light signals or visual markings indicating that turns should be made to the right. (PLT435) — 14 CFR §91.126

Answers (A) and (C) are incorrect because the traffic pattern altitude will vary for each airport and aircraft; however, all turns should be to the left unless otherwise indicated.

ALL
2002. Entries into traffic patterns while descending create specific collision hazards and

A—should be avoided.
B—should be used whenever possible.
C—are illegal.

Entries into traffic patterns while descending create specific collision hazards and should be avoided. (PLT170) — AIM ¶4-4-15

AIR, RTC, WSC, PPC
2324. (Refer to Figure 64.) The traffic patterns indicated in the segmented circle have been arranged to avoid flights over an area to the

A—south of the airport.
B—north of the airport.
C—southeast of the airport.

The segmented circle depicts the direction of landing for each airport; the long part of the "L" depicts the runway, and the short part depicts the airport traffic pattern for that runway. This figure indicates there are no flights that cross the southeast area of the airport. (PLT039) — AIM ¶4-3-4

Answers

2194 [B]	2002 [A]	2324 [C]

Airport Markings and Signs

Both the *Aeronautical Information Manual* (AIM) and the *Pilot's Handbook of Aeronautical Knowledge* (FAA-H-8083-25) contain explanations of airport signs and markings.

Looking at FAA Figure 72 in the Computer Testing Supplement:

- **A** is a surface painted runway marking.
- **B** is a stop bar/ILS hold.
- **C** is the vehicle lanes.
- **D** is a hold marking for land and hold short operations.
- **E** is a taxiway/taxiway hold marking.
- **F** is a taxiway edge marking (do not cross).

Looking at FAA Figure 71 in the Computer Testing Supplement:

- **A** is a taxiway/runway hold position sign.
- **B** is a runway approach hold position sign.
- **C** is an ILS critical area hold position sign.
- **D** is a no entry sign.
- **E** is a taxiway location sign.
- **F** is a runway location sign.
- **G** is a runway safety area/obstacle free zone boundary.
- **H** is an ILS critical area boundary.
- **I** is an inbound destination sign.
- **J** is an outbound destination sign.
- **K** is a taxiway direction sign.
- **L** is a runway distance remaining (in 1,000 foot increments).
- **M** is a runway/runway hold position sign.
- **N** is a taxiway ending marker.

A closed runway which is unusable and may be hazardous, even though it may appear usable, will be marked by an "X"

ALL

2206. Which publication contains an explanation of airport signs and markings?

A—Aeronautical Information Manual (AIM).
B—Advisory Circulars (AC).
C—Airport Facility Directory (A/FD).

Both the Aeronautical Information Manual *(AIM) and the* Pilot's Handbook of Aeronautical Knowledge *(FAA-H-8083-25) contain explanations of airport signs and markings. (PLT141) — AIM Chapter 2, Section 3*

Answer (B) is incorrect because ACs provide information on nonregulatory material of interest. Answer (C) is incorrect because the A/FD provides only comprehensive information on a given airport (such as runway lengths, available services, lighting, etc.).

Answers
2206 [A]

ALL

2021. (Refer to Figure 62.) That portion of the runway identified by the letter A may be used for

A—landing.
B—taxiing and takeoff.
C—taxiing and landing.

Thresholds are marked at the beginning of a full-strength runway surface able to endure landing impacts or at a point on the runway which will encourage pilots to avoid short approaches due to hidden noise or obstacle problems. Area A of FAA Figure 62 is marked with arrows which point towards a displaced threshold. Thus, the paved surface prior to the threshold is available for taxi, takeoff and landing rollout, but not for touchdown. (PLT141) — AIM ¶2-3-3

Answers (A) and (C) are incorrect because area A may be used for the landing rollout, not the actual landing or point of touchdown.

ALL

2022. (Refer to Figure 72) Which marking indicates a vehicle lane?

A—A
B—C
C—E

The vehicle roadway markings are used when necessary to define a pathway for vehicle operations on or crossing areas that are also intended for aircraft. These markings consist of a white solid line to delineate each edge of the roadway and a dashed line to separate lanes within the edges of the roadway. In lieu of the solid lines, zipper markings may be used to delineate the edges of the vehicle roadway. (PLT141) — AIM ¶2-3-6

Answer (A) is incorrect because Area A is a surface painted holding position sign. Answer (C) is incorrect because Area E is the taxiway/taxiway hold marking.

ALL

2023. (Refer to Figure 63.) The arrows that appear on the end of the north/south runway indicate that the area

A—may be used only for taxiing.
B—is usable for taxiing, takeoff, and landing.
C—cannot be used for landing, but may be used for taxiing and takeoff.

The paved area behind the displaced runway threshold is available for taxiing, landing rollout, and the takeoff of aircraft. (PLT141) — AIM ¶2-3-3

Answer (A) is incorrect because this area may be used for both taxiing and takeoff. Answer (B) is incorrect because this area may be used for the landing rollout; not the actual landing (touchdown).

ALL

2024. The numbers 35 and 17 on a runway indicate that the runway is oriented approximately

A—035°; and 017°; magnetic heading.
B—350°; and 170°; magnetic heading.
C—350°; and 170°; true heading.

The runway number is the whole number nearest one-tenth the magnetic azimuth of the centerline of the runway, measured clockwise from magnetic north. For example: 352° = RWY 35; 172° = RWY 17. (PLT141) — AIM ¶2-3-3

Answer (A) is incorrect because runway numbers drop the last digit from the heading. Answer (C) is incorrect because runway numbers are based on magnetic direction.

ALL

2025. 'Runway hold position' markings on the taxiway

A—identify areas where aircraft are prohibited.
B—identify where aircraft hold short of the runway.
C—allow an aircraft permission onto the runway.

Runway holding position markings on taxiways identify the locations on a taxiway where an aircraft is supposed to stop when it does not have clearance to proceed onto the runway. (PLT141) — AIM ¶2-3-5

ALL

2026. The 'runway hold position' sign denotes

A—an entrance to a runway from a taxiway.
B—an area protected for an aircraft approaching a runway.
C—an entrance to a taxiway from a runway.

Mandatory instruction signs are used to denote an entrance to a runway or critical area and areas where an aircraft is prohibited from entering. The runway holding position sign is located at the holding position on taxiways that intersect a runway or on runways that intersect other runways. (PLT141) — AIM ¶2-3-8

Answers

2021 [B]	2022 [B]	2023 [C]	2024 [B]	2025 [B]	2026 [A]

ALL
2027. What is the purpose of the runway hold position sign?

A—Denotes area protected for an aircraft approaching or departing a runway.
B—Denotes runways that intersect other runways.
C—Denotes an entrance to taxiway from a runway.

Mandatory instruction signs are used to denote an entrance to a runway or critical area and areas where an aircraft is prohibited from entering. The runway holding position sign is located at the holding position on taxiways that intersect a runway or on runways that intersect other runways. (PLT141) — AIM ¶2-3-8

ALL
2028. What is the purpose for the runway hold position markings on the taxiway?

A—Holds aircraft short of the runway.
B—Allows an aircraft permission onto the runway.
C—Identifies area where aircraft are prohibited.

Runway holding position markings on taxiways identify the locations on a taxiway where an aircraft is supposed to stop when it does not have clearance to proceed onto the runway. (PLT141) — AIM ¶2-3-5

ALL
2029. (Refer to Figure 71) Which sign indicates the runway on which the aircraft is located?

A—E
B—F
C—L

"F" is a runway location sign, which identifies the runway on which the aircraft is located. (PLT141) — AIM ¶2-3-9

Answer (A) is incorrect because E is a taxiway location sign, which identifies the taxiway on which the aircraft is located. Answer (C) is incorrect because this is a runway distance remaining sign, which identifies the runway length remaining.

ALL
2030. Holding position signs have

A—red inscriptions on white background.
B—white inscriptions on red background.
C—yellow inscriptions on red background.

Hold position signs have white inscriptions on red background. (PLT141) — AIM ¶2-3-5

Answers (A) and (C) are incorrect because these color combination are not used in airport signs.

ALL
2031. 'Runway hold position' markings on the taxiway

A—identifies where aircraft hold short of the runway.
B—identifies an area where aircraft are prohibited.
C—allows an aircraft permission onto the runway.

Runway holding position markings on taxiways identify the locations on a taxiway where an aircraft is supposed to stop when it does not have clearance to proceed onto the runway. (PLT141) — AIM ¶2-3-5

ALL
2032. (Refer to Figure 72) Which symbol indicates a taxiway/taxiway intersection hold position marking?

A—B
B—D
C—E

Symbol "E" indicates a taxiway/taxiway hold marking. (PLT141) — AIM ¶2-3-4

Answer (A) is incorrect because B is a stop bar/ILS hold marking. Answer (B) is incorrect because D is a hold marking for land and hold short operations.

ALL
2218. The "taxiway ending" marker

A—indicates taxiway does not continue.
B—identifies area where aircraft are prohibited.
C—provides general taxiing direction to named taxiway.

Taxiway ending markers are used to indicate that the taxiway does not continue. (PLT141) — AIM ¶2-3-4

ALL
2219. What is the purpose of No Entry sign?

A—Identifies paved area where aircraft are prohibited from entering.
B—Identifies area that does not continue beyond intersection.
C—Identifies the exit boundary for the runway protected area.

The "no entry" sign prohibits an aircraft from entering an area. Typically, this sign would be located on a taxiway intended to be used in only one direction or at the intersection of vehicle roadways with runways, taxiways or aprons where the roadway may be mistaken as a taxiway or other aircraft movement surface. (PLT141) — AIM ¶2-3-8

Answer (B) is incorrect because this is the purpose of a hold position sign. Answer (C) is incorrect because this is the purpose of the runway boundary sign.

Answers

2027 [B]	2028 [A]	2029 [B]	2030 [B]	2031 [A]	2032 [C]
2218 [A]	2219 [A]				

Airport Lighting

At night, the location of an airport can be determined by the presence of an airport **rotating beacon** light. The colors and color combinations that denote the type of airports are:

White and green .... Lighted land airport
*Green alone .... Lighted land airport
White and yellow .... Lighted water airport
*Yellow alone .... Lighted water airport
Green, yellow, white .... Lighted heliport

Note: Green alone or amber alone is used only in connection with a white-and-green or white-and-amber beacon display, respectively.

A civil-lighted land airport beacon will show alternating white and green flashes. A military airfield will be identified by dual-peaked (two quick) white flashes between green flashes.

In Class B, C, D, or E airspace, operation of the airport beacon during the hours of daylight often indicates the ceiling is less than 1,000 feet and/or the visibility is less than 3 miles. However, pilots should not rely solely on the operation of the airport beacon to indicate if weather conditions are IFR or VFR.

Runway edge lights are used to outline the runway at night or during periods of low visibility. For the most part, runway edge lights are white, and may be high-, medium-, or low-intensity, while taxiways are outlined by blue omnidirectional lights.

Radio control of lighting is available at some airports, providing airborne control of lights by keying the aircraft's microphone. The control system is responsive to 7, 5, or 3 microphone clicks. Keying the microphone 7 times within 5 seconds will turn the lighting to its highest intensity; 5 times in 5 seconds will set the lights to medium intensity; low intensity is set by keying 3 times in 5 seconds.

ALL

2019. An airport's rotating beacon operated during daylight hours often indicates

A—there are obstructions on the airport.
B—that weather at the airport located in Class D airspace is below basic VFR weather minimums.
C—the Air Traffic Control tower is not in operation.

In Class B, C, D or E airspace, operation of the airport beacon during the hours of daylight often indicates that the weather in the airspace is below basic VFR weather minimums (ground visibility is less than 3 miles and/or the ceiling is less than 1,000 feet). (PLT116) — AIM ¶2-1-8

Answer (A) is incorrect because obstructions are noted in the A/FD or NOTAMs. Answer (C) is incorrect because tower operation is noted in the A/FD and on the sectional, not with a special light signal.

Answers
2019 [B]

Visual Approach Slope Indicator (VASI)

The **Visual Approach Slope Indicator (VASI)** is a lighting system arranged so as to provide visual-descent guidance information during approach to a runway. The lights are visible for up to 5 miles during the day. The VASI glide path provides obstruction clearance, while lateral guidance is provided by the runway or runway lights. When operating to an airport with an operating control tower, the pilot of an airplane approaching to land on a runway served by a VASI is required to maintain an altitude at or above the glide slope until a lower altitude is necessary for landing.

Most VASI installations consist of two bars, near and far, which provide one visual glide path. On final approach flying toward the runway of intended landing, if the pilot sees both bars as red, the aircraft is below the glide path (Figure 5-8A). Maintaining altitude, the pilot will see the near bar turn pink and then white, while the far bar remains red, indicating the glide path is being intercepted (Figure 5-8B). If the aircraft is above the glide path, the pilot will see both near and far bars as white (Figure 5-8C).

Tri-color visual approach slope indicators normally consist of a single light unit projecting a three-color visual approach path. The below-glide path indication is red, the above-glide path indication is amber, and the on-glide path indication is green. *See* Figure 5-9.

Pulsating VASIs normally consist of a single light unit projecting a two-color visual approach path. The below-glide path indication is normally red or pulsating red, and the above-glide path indication is normally pulsating white. The on-glide path indication is usually steady white. *See* Figure 5-10.

The **Precision Approach Path Indicator (PAPI)** uses a single row of lights. Four white lights means "too high." One red light and three white lights means "slightly high," etc. *See* Figure 5-11.

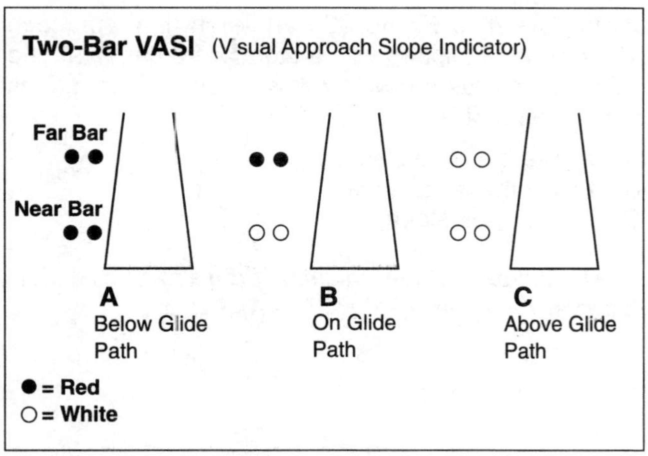

Figure 5-8. A 2-bar VASI

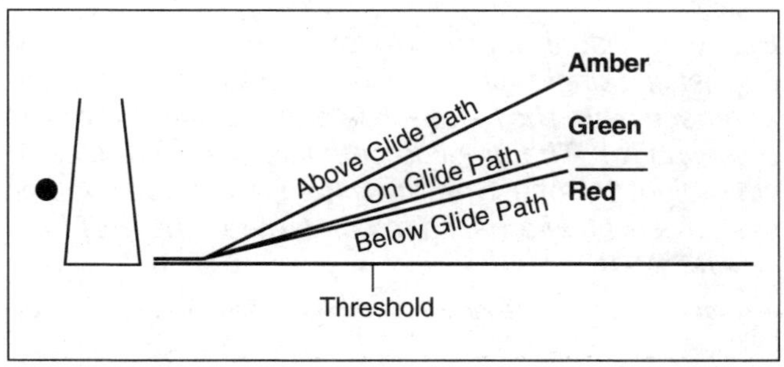

Figure 5-9. Tri-color VASI

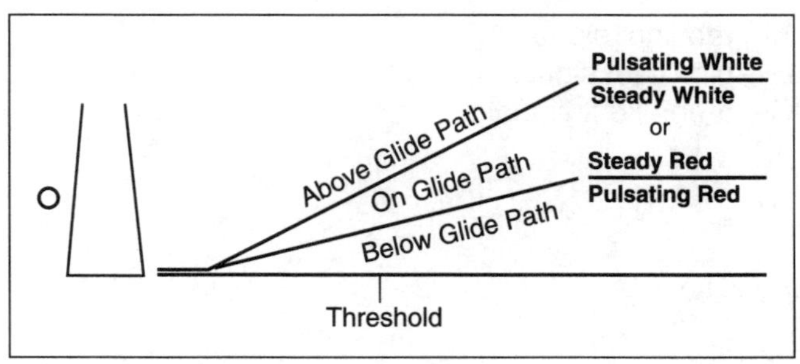

Figure 5-10. Pulsating VASI system

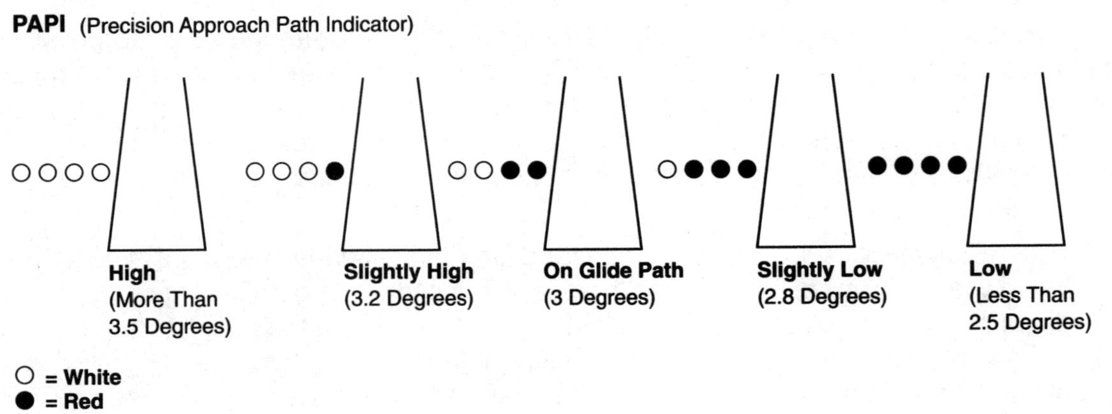

Figure 5-11. Precision approach path indicator (PAPI)

ALL

2018. A below glide slope indication from a pulsating visual approach slope indicator is a

A—pulsating white light.
B—steady white light.
C—pulsating red light.

Pulsating visual approach slope indicators normally consist of a single light unit projecting a two-color visual approach path. The below-glide path indication is red or pulsating red. The on-glide path indication is a steady white light for one type of system, while for another system it is an alternating red and white light. (PLT147) — AIM ¶2-1-2

Answers (A) and (B) are incorrect because white indicates on-glide path.

AIR, RTC, WSC, PPC

2282. A below glide slope indication from a tri-color VASI is a

A—red light signal.
B—pink light signal.
C—green light signal.

A tri-color VASI normally is a single light unit projecting a three-color visual approach path. Below the glide path is red, on the glide path is green, and above the glide path is amber. (PLT147) — AIM ¶2-1-2

AIR, RTC, WSC, PPC

2325. (Refer to Figure 65.) While on final approach to a runway equipped with a standard 2-bar VASI, the lights appear as shown by illustration D. This means that the aircraft is

A—above the glide slope.
B—below the glide slope.
C—on the glide slope.

The below-glide slope indication from a two-bar VASI is red over red lights. (PLT147) — AIM ¶2-1-2

Answers

2018 [C] 2282 [A] 2325 [B]

Airspace and Charts

Controlled airspace, that is, airspace within which some or all aircraft may be subject to air traffic control, consists of those areas designated as Class A, Class B, Class C, Class D, and Class E airspace.

FAA Computer Testing Supplement Legend 1 is on every Sectional Chart and provides the key to understand and identify the airspace areas and chart symbols.

Much of the Class E controlled airspace begins at either 700 feet or 1,200 feet above the ground. Most of the unmarked airspace on a sectional is Class E airspace starting at 1,200 feet AGL and extending up to 18,000 feet MSL. When the floor of Class E airspace drops to 700 feet AGL, it is indicated by a magenta vignette surrounding most large and small airports. The lateral limits and floors of Class E airspace of 700 feet are defined by a magenta vignette; while the lateral limits and floors of 1,200 feet are defined by a blue vignette if it abuts uncontrolled airspace. Floors other than 700 feet or 1,200 feet are indicated by a number indicating the floor.

Sport Pilots are allowed to fly in Class E and G airspace but must obtain additional training and receive a logbook endorsement to fly into Class B, C, and D airspace. *See* Figure 5-12. This endorsement is also required for flying at any airport where a control tower is in operation. Some Class B airports (identified in 14 CFR Part 91 Appendix D) are restricted unless you hold at least a Private Pilot certificate. Sport Pilots are required to maintain 3 miles visibility in all airspace. No pilot may operate an aircraft under VFR at airports in Class B, C, or D airspace when the ceiling is less than 1,000 feet.

Class A—Class A airspace extends from 18,000 feet MSL up to and including FL600 and is not depicted on VFR sectional charts. No flight under visual flight rules (VFR), including VFR-On-Top, is authorized in Class A airspace. Class A airspace is of little concern to Sport Pilots because they may not fly above 10,000 feet MSL.

Continued

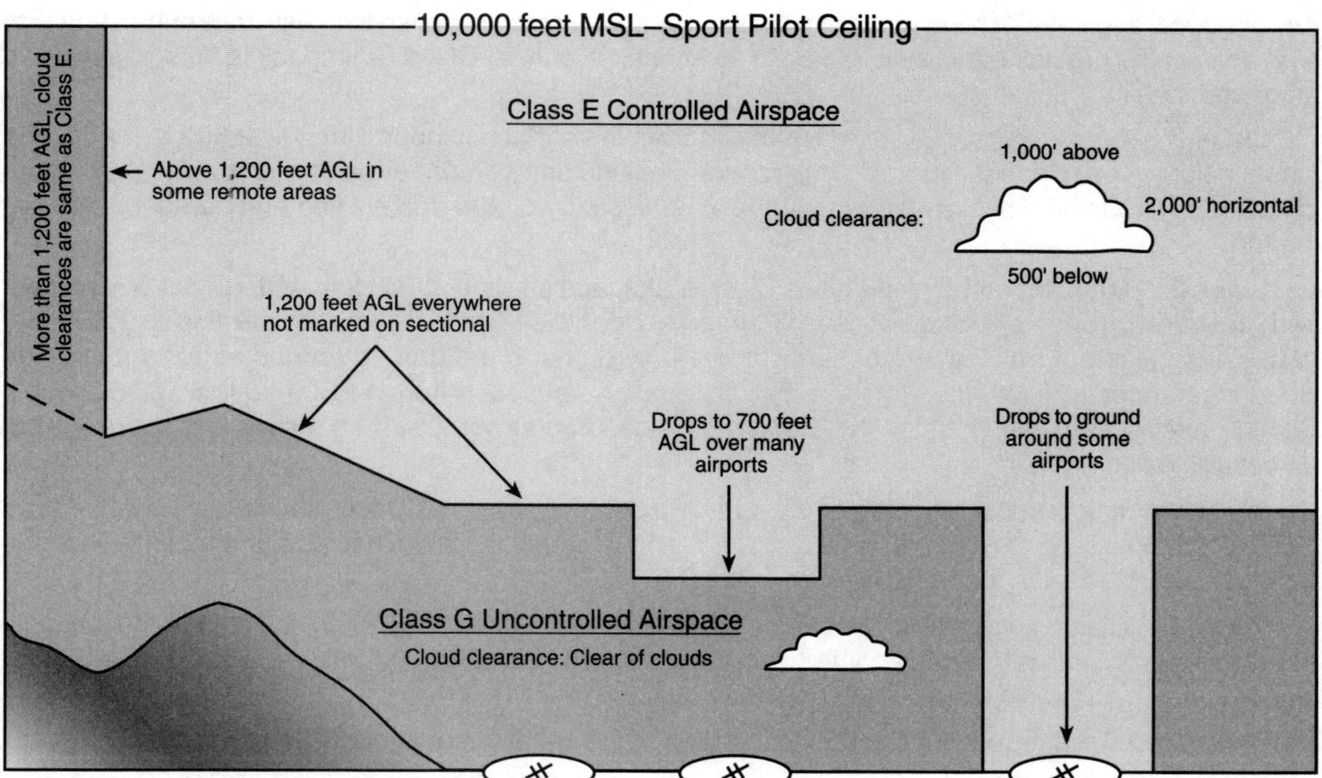

Figure 5-12. Sport pilot airspace

Class B—Class B airspace consists of controlled airspace extending upward from the surface or higher to specified altitudes. Each Class B airspace sector, outlined in blue on the sectional aeronautical chart, is labeled with its delimiting altitudes. On the Terminal Area Chart, each Class B airspace sector is, again, outlined in blue and is labeled with its delimiting arcs, radials, and altitudes. Each Class B airspace location will contain at least one primary airport. An ATC clearance is required prior to operating within Class B airspace. A pilot landing or taking off from one of a group of 12 specific, busy airports must hold at least a Private Pilot Certificate. A student pilot may not operate an aircraft on a solo flight within Class B airspace or to, from, or at an airport located within Class B airspace unless both ground and flight instruction has been received from an authorized instructor to operate within that Class B airspace or at that airport, and the flight and ground instruction has been received within that Class B airspace or at the specific airport for which the solo flight is authorized. The student's logbook must be endorsed within the preceding 90 days by the instructor who gave the flight training and the endorsement must specify that the student has been found competent to conduct solo flight operations in that Class B airspace or at that specific airport. Each airplane operating within Class B airspace must be equipped with a two-way radio with appropriate ATC frequencies, and a 4096 code transponder with Mode C automatic altitude-reporting capability.

Class C—All Class C airspace has the same dimensions with minor site variations. They are composed of two circles both centered on the primary airport. The surface area has a radius of 5 nautical miles and extends from the surface up to 4,000 feet above the airport. The shelf area has a radius of 10 nautical miles and extends vertically from 1,200 feet AGL up to 4,000 feet above the primary airport. In addition to the Class C airspace proper, there is an outer area with a radius of 20 nautical miles and vertical coverage from the lower limits of the radio/radar coverage up to the top of the approach control facility's delegated airspace. Within the outer area, pilots are encouraged to participate but it is not a VFR requirement. Class C airspace service to aircraft proceeding to a satellite airport will be terminated at a sufficient distance to allow time to change to the appropriate tower or advisory frequency. Aircraft departing satellite airports within Class C airspace shall establish two-way communication with ATC as soon as practicable after takeoff. On aeronautical charts, Class C airspace is depicted by solid magenta lines.

Class D—Class D airspace extends upward from the surface to approximately 2,500 feet AGL (the actual height is as needed). Class D airspace may include one or more airports and is normally 4 nautical miles in radius. The actual size and shape is depicted by a blue dashed line and numbers showing the top.

Class E—Magenta shading identifies Class E airspace starting at 700 feet AGL (Class G is below), and no shading (or blue if next to Class G airspace) identifies Class E airspace starting at 1,200 feet AGL (Class G is below). Class E airspace commonly drops to the ground around airports to provide cloud clearances. It may also start at other altitudes. All airspace from 14,500 feet to 17,999 feet is Class E airspace. It also includes the surface area of some airports with an instrument approach but no control tower.

An **airway** is a corridor of controlled airspace extending from 1,200 feet above the surface (or as designated) up to and including 17,999 feet MSL, and 4 nautical miles either side of the centerline. The airway is indicated by a centerline, shown in blue.

Class G—Class G airspace is airspace within which Air Traffic Control (ATC) has neither the authority nor responsibility to exercise any control over air traffic, and pilots can fly with ceilings less than 1,000 feet AGL and clear of clouds.

Prohibited Areas are blocks of airspace within which the flight of aircraft is prohibited.

Restricted Areas denote the presence of unusual, often invisible, hazards to aircraft such as artillery firing, aerial gunnery, or guided missiles. Penetration of Restricted Areas without authorization of the using or controlling agency may be extremely hazardous to the aircraft and its occupants.

Warning Areas contain the same hazardous activities as those found in Restricted Areas, but are located in international airspace. Prohibited, restricted, or warning areas are depicted as shown in FAA Legend 1.

Military Operations Areas (MOAs) consist of airspace established for the purpose of separating certain military training activities from instrument flight rules (IFR) traffic. Pilots operating under VFR should exercise extreme caution while flying within an active MOA. Any Flight Service Station (FSS) within 100 miles of the area will provide information concerning MOA hours of operation. Prior to entering an active MOA, pilots should contact the controlling agency for traffic advisories.

Alert Areas may contain a high volume of pilot training activities or an unusual type of aerial activity, neither of which is hazardous to aircraft. Pilots of participating aircraft as well as pilots transiting the area are equally responsible for collision avoidance.

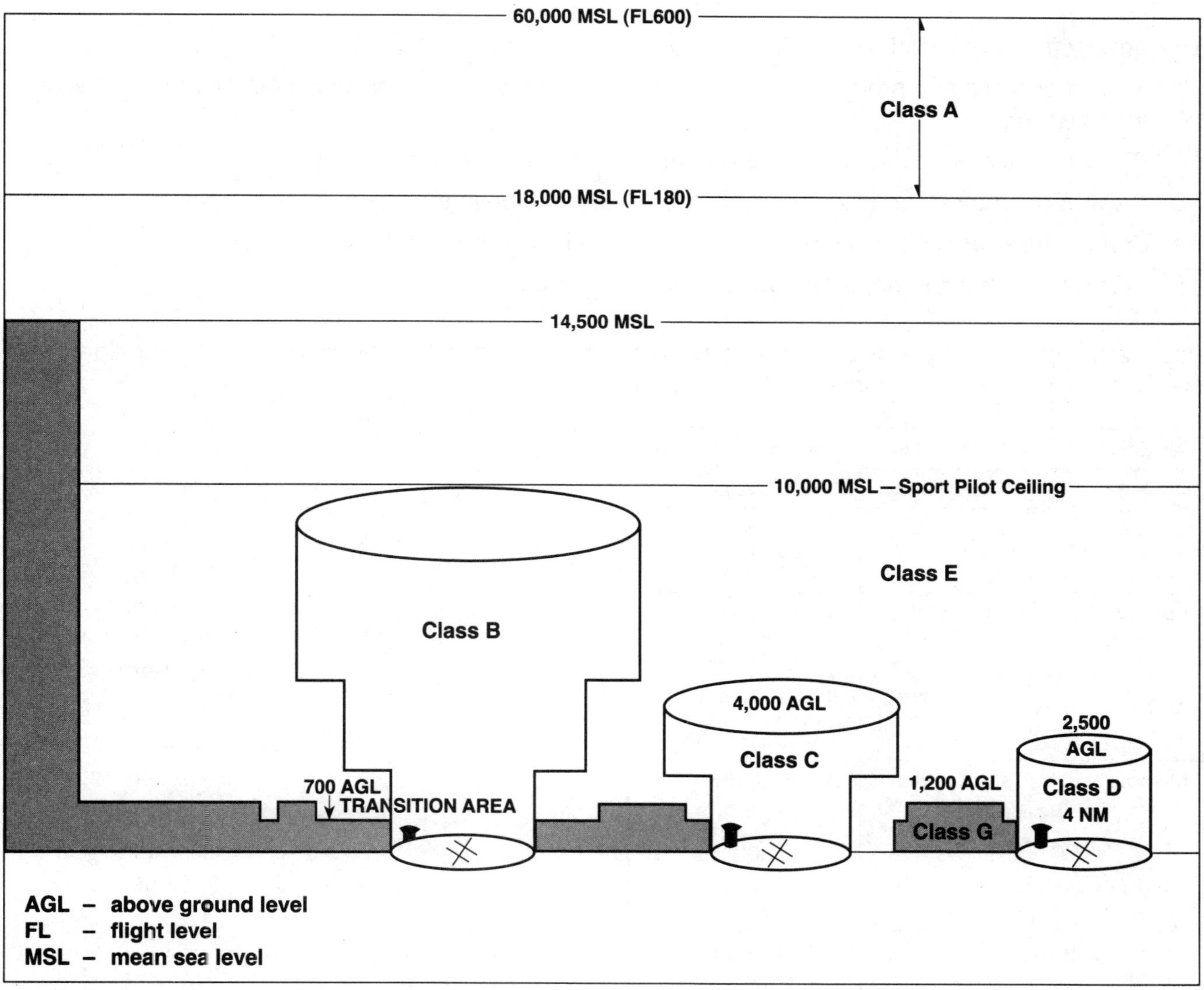

Figure 5-13. Airspace around control-towered airports

An **Airport Advisory Area** is the area within 10 statute miles of an airport where a control tower is not in operation but where a Flight Service Station (FSS) is located. The FSS provides advisory service to aircraft arriving and departing. It is not mandatory for pilots to use the advisory service, but it is strongly recommended that they do so.

Aircraft are requested to remain at least 2,000 feet above the surface of National Parks, National Monuments, Wilderness and Primitive Areas, and National Wildlife Refuges.

Military Training Routes (MTRs) have been developed for use by the military for the purpose of conducting low-altitude, high-speed training. Generally, MTRs are established below 10,000 feet MSL for operations at speeds in excess of 250 knots.

IFR Military Training Routes (IR) operations are conducted in accordance with instrument flight rules (IFRs), regardless of weather conditions. **VFR Military Training Routes (VR)** operations are conducted in accordance with visual flight rules (VFRs). IR and VR at and below 1,500 feet AGL (with no segment above 1,500) will be identified by four digit numbers, e.g., VR1351, IR1007. IR and VR above and below 1,500 feet AGL (segments of these routes may be below 1,500) will be identified by three digit numbers, e.g., IR341, VR426.

Temporary flight restrictions (TFR) are imposed in order to:

1. Protect persons and property in the air or on the surface from an existing or imminent flight associated hazard;
2. Provide a safe environment for the operation of disaster relief aircraft;
3. Prevent an unsafe congestion of sightseeing aircraft above an incident;
4. Protect the President, Vice President, or other public figures; and,
5. Provide a safe environment for space agency operations.

Pilots are expected to check appropriate NOTAMs during flight planning when conducting flight in an area where a temporary flight restriction is in effect.

Airspace	Flight visibility	Distance from clouds
Class B	3 statute miles	Clear of clouds.
Class C	3 statute miles	500 feet below. 1,000 feet above. 2,000 feet horizontal.
Class D	3 statute miles	500 feet below. 1,000 feet above. 2,000 feet horizontal.
Class E	3 statute miles	500 feet below. 1,000 feet above. 2,000 feet horizontal.
Class G: 1,200 feet or less above the surface (regardless of MSL altitude)	3 statute miles	Clear of clouds.
More than 1,200 feet above the surface.	3 statute miles	500 feet below. 1,000 feet above. 2,000 feet horizontal.

Figure 5-14. Basic weather minimums for Sport Pilots

ALL
2041. (Refer to Figure 57, area 2.) The floor of Class B airspace at Addison Airport is

A—3,000 feet MSL.
B—at the surface.
C—3,100 feet MSL.

The thick blue lines on the sectional chart indicate the boundaries of the overlying Class B airspace. Within each segment, the floor and ceiling are denoted by one number over a second number or the letters SFC. The floor of the Class B airspace is 3,000 feet MSL. (PLT040) — AIM ¶3-2-3

ALL
2042. (Refer to Figure 57, area 4.) The floor of Class B airspace overlying Hicks Airport (T67) north-northwest of Fort Worth Meacham Field is

A—at the surface.
B—3,200 feet MSL.
C—4,000 feet MSL.

The thick blue lines on the sectional chart indicate the boundaries of the overlying Class B airspace. Within each segment, the floor and ceiling are denoted by one number over a second number or the letters SFC. The floor of the Class B airspace is 4,000 feet MSL. (PLT040) — AIM ¶3-2-3

ALL
2304. (Refer to Figure 57, area 4.) The airspace directly overlying Fort Worth Meacham is

A—Class B airspace to 10,000 feet MSL.
B—Class C airspace to 5,000 feet MSL.
C—Class D airspace to 3,200 feet MSL.

The dashed blue lines surrounding Fort Worth Meacham (FTW) indicate Class D airspace; [32] indicates the ceiling of the Class D airspace in hundreds of feet. (PLT064) — Sectional Chart Legend

ALL
2305. (Refer to Figure 59, area 6.) The airspace overlying and within 5 miles of Barnes County Airport is

A—Class D airspace from the surface to the floor of the overlying Class E airspace.
B—Class E airspace from the surface to 1,200 feet MSL.
C—Class G airspace from the surface to 700 feet AGL.

The magenta shading around the airport indicate Class E airspace starting at 700 feet AGL; Class G airspace below this. (PLT064) — Sectional Chart Legend

Answer (A) is incorrect because Class D airspace would be depicted with blue dashed lines. Answer (B) is incorrect because this would be depicted with blue shading around the airport.

ALL
2192. Which is true regarding flight operations in Class B airspace?

A—The pilot must receive an ATC clearance before operating an aircraft in that area.
B—Flight under VFR is not authorized unless the pilot in command is instrument rated.
C—Solo student pilot operations are not authorized.

No person may operate an aircraft within Class B airspace unless the operator first receives an ATC clearance from the ATC facility having jurisdiction for that area. (PLT162) — 14 CFR §91.131

Answer (B) is incorrect because a Private pilot certificate without an instrument rating is sufficient to operate in Class B airspace. Answer (C) is incorrect because with the proper training and endorsements, a student pilot may operate in Class B airspace.

Answers

2041 [A]	2042 [C]	2304 [C]	2305 [C]	2192 [A]

ALL
2193. Which is true regarding flight operations in Class B airspace?

A—The pilot in command must hold at least a private pilot certificate with an instrument rating.
B—The pilot in command must hold at least a student pilot certificate.
C—The aircraft must be equipped with an ATC transponder and altitude reporting equipment.

You may not operate an aircraft in Class B airspace unless the aircraft is equipped with an operating transponder and automatic altitude reporting equipment or you have requested a waiver at least one hour before the proposed operation. Both 14 CFR §§91.130 and 91.131 have provisions to allow any aircraft to operate in the airspace without transponders. (PLT162) — 14 CFR §91.131

Answer (A) is incorrect because a private pilot certificate without an instrument rating is sufficient to operate in Class B airspace. Answer (B) is incorrect because Class B airspace requires the pilot-in-command hold a private pilot certificate or have a logbook endorsement confirming that the necessary training was received.

ALL
2043. Which is true regarding flight operations to a satellite airport, without an operating control tower, within Class C airspace?

A—Prior to entering that airspace, a Sport Pilot must contact the FSS.
B—Prior to entering that airspace, a Sport Pilot must contact the primary airport tower.
C—Prior to entering that airspace, a Sport Pilot must receive the appropriate logbook endorsement.

A Sport Pilot must receive ground training and a logbook endorsement to operate in Class B, C, and D airspace. (PLT162) — 14 CFR §61.94

Answer (A) is incorrect because airports in Class C airspace have a control tower; all communications must be established with this facility, not FSS. Answer (B) is incorrect because the Sport Pilot must first have specific ground training and a logbook endorsement before attempting to enter Class C airspace.

ALL
2200. Which initial action should a pilot take prior to entering Class C airspace?

A—Contact approach control on the appropriate frequency.
B—Contact the tower and request permission to enter.
C—Contact the Flight Service Station (FSS) for traffic advisories.

Radio contact is required to operate in Class C airspace, but permission is not required. (PLT161) — AIM ¶3-2-4

ALL
2044. The vertical limit of Class C airspace above the primary airport is normally

A—1,200 feet AGL.
B—3,000 feet AGL.
C—4,000 feet AGL.

Class C airspace consists of two circles, both centered on the primary/Class C airspace airport. The surface area has a radius of 5 NM. The shelf area usually has a radius of 10 NM. The airspace of the surface area usually extends from the surface of the Class C airspace airport up to 4,000 feet above that airport. The airspace area between the 5 and 10 NM rings usually begins at a height of 1,200 feet AGL and extends to the same altitude cap as the inner circle. These dimensions may be varied to meet individual situations. (PLT161) — AIM ¶3-2-4

ALL
2045. (Refer to Figure 69, area 3.) What is the floor of the Savannah Class C airspace at the shelf area (outer circle)?

A—1,300 feet AGL.
B—1,300 feet MSL.
C—1,700 feet MSL.

Within the outer magenta circle of Savannah Class C airspace, there is a number 41 directly above the number 13. These depict the floor and ceiling of the Class C airspace; the floor being 1,300 feet MSL and the ceiling being 4,100 feet MSL. (PLT040) — AIM ¶3-2-4

Answers

2193 [C]	2043 [C]	2200 [A]	2044 [C]	2045 [B]

ALL

2046. The normal radius of the outer area of Class C airspace is

A—5 nautical miles.
B—15 nautical miles.
C—20 nautical miles.

The normal radius of the outer area will be 20 NM. This is the area where separation is provided after two-way communication is established. Communications with ATC are required before you enter the 10 NM shelf area inside the Class C airspace; however, communications are not required while you are in this 20 NM Class C "outer area." (PLT161) — AIM ¶3-2-4

ALL

2047. A blue segmented circle on a Sectional Chart depicts which class airspace?

A—Class B.
B—Class C.
C—Class D.

A blue segmented circle on a sectional chart depicts Class D airspace which means a control tower is present. (PLT064) — 14 CFR §1.1

Answer (A) is incorrect because Class B airspace is depicted by a solid blue line. Answer (B) is incorrect because Class C airspace is depicted by a solid magenta line.

ALL

2048. Airspace at an airport with a part-time control tower is classified as Class D airspace only

A—when the prevailing visibility is below 3 statute miles.
B—when the associated control tower is in operation.
C—when the associated Flight Service Station is in operation.

Class D airspace means a control tower is in operation. If the tower closes, it reverts to Class E airspace. (PLT161) — 14 CFR §1.1

Answer (A) is incorrect because the visibility does not affect the airspace classification. Answer (C) is incorrect because an FSS has nothing to do with the class of airspace.

ALL

2049. What designated airspace associated with an airport becomes inactive when the control tower at that airport is not in operation?

A—Class D, which then becomes Class C.
B—Class D, which then becomes Class E.
C—Class B.

Class D airspace means a control tower is in operation. If the tower closes, it reverts to Class E airspace. (PLT161) — 14 CFR §1.1

ALL

2241. Unless otherwise specified, Federal Airways include that Class E airspace extending upward from

A—700 feet above the surface up to and including 17,999 feet MSL.
B—1,200 feet above the surface up to and including 17,999 feet MSL.
C—the surface up to and including 18,000 feet MSL.

Federal airways are part of Class B, C, D, or E airspace. They are 8 miles wide, 4 miles either side of centerline. They usually begin at 1,200 feet AGL and continue up to but not including 18,000 feet MSL, or FL180. (PLT161) — 14 CFR §71.5

ALL

2050. (Refer to Figure 70, point 1) What minimum altitude is required to avoid the Livermore Airport (LVK) Class D airspace?

A—2,503 feet MSL.
B—2,901 feet MSL.
C—3,297 feet MSL.

The Class D airspace at Livermore has a top of 2,900 feet MSL, indicated by the [29] within the blue segmented circle. Therefore, the minimum altitude to fly over and avoid the Class D airspace is 2,901 feet MSL. (PLT040) — AIM ¶3-2-5

Answer (A) is incorrect because 2,503 feet MSL would place you within the Class D airspace. Answer (C) is incorrect because although 3,297 feet MSL would keep you outside the Class D airspace, it is not the minimum altitude required to avoid it.

Answers

2046 [C]	2047 [C]	2048 [B]	2049 [B]	2241 [B]	2050 [B]

ALL
2051. (Refer to Figure 66, area 3.) The base of that portion of Class E airspace designated as a Federal Airway over Magee Airport are

A—1,200 feet AGL.
B—700 feet MSL.
C—7,500 feet MSL.

The VOR and L/MF Airway System consists of airways designated from 1,200 feet above the surface (in some instances higher) up to, but not including, 18,000 feet MSL. (PLT064) — AIM ¶3-2-6

ALL
2052. The purpose of Military Training Routes, charted as VFR Military Training Routes (VR) and IFR Military Training Routes (IR) on sectional charts, is to ensure the greatest practical level of safety for all flight operations and to allow the military to conduct

A—low altitude, high-speed training.
B—radar instrument training.
C—air-to-air refueling training.

The military services must train in a wide range of airborne tactics. One phase of this training involves "low level" combat tactics. The required maneuvers and high speeds are such that they may occasionally make the see-and-avoid aspect of VFR flight more difficult without increased vigilance in areas containing such operations. In an effort to ensure the greatest practical level of safety for all flight operations, the Military Training Route (MTR) program was conceived. (PLT116) — AIM ¶3-5-2

ALL
2055. (Refer to Figure 59, area 2.) What hazards to aircraft may exist in areas such as Devils Lake East MOA?

A—Unusual, often invisible, hazards to aircraft such as artillery firing, aerial gunnery, or guided missiles.
B—Military activities including, air combat tactics, aerobatics and low-altitude tactics.
C—High volume of pilot training or an unusual type of aerial activity.

A Military Operations Area (MOA) contains military training activities such as aerobatics, and calls for extreme caution. (PLT393) — AIM ¶3-4-5

ALL
2056. Who is responsible for collision avoidance in a Military Operations Area (MOA)?

A—Each pilot.
B—ATC controllers.
C—Military controllers.

Pilots operating under VFR should exercise extreme caution while flying within a MOA when military activity is being conducted. No clearance is necessary to enter a MOA. (PLT162) — AIM ¶3-4-5

ALL
2057. (Refer to Figure 56 area 4.) What hazards to aircraft may exist in restricted areas such as R-5302B?

A—Unusual, often invisible, hazards such as aerial gunnery or guided missiles.
B—Military training activities that necessitate acrobatic or abrupt flight maneuvers.
C—High volume of pilot training or an unusual type of aerial activity.

Restricted Areas denote the presence of unusual, often invisible, hazards to aircraft such as artillery firing, aerial gunnery, or guided missiles. Penetration of Restricted Areas without authorization of the using or controlling agency may be extremely hazardous to the aircraft and its occupants. (PLT064) — AIM ¶3-4-3

Answer (B) is incorrect because this defines a military operations area. Answer (C) is incorrect because this defines an alert area.

ALL
2058. (Refer to Figure 59, area 3.) When flying over Arrowwood National Wildlife Refuge, a pilot should fly no lower than

A—2,000 feet AGL.
B—2,500 feet AGL.
C—3,000 feet AGL.

All aircraft are requested to maintain a minimum altitude of 2,000 feet above the surface of national parks, monuments, seashores, lakeshores, recreation areas, and scenic riverways administered by the National Park Service, National Wildlife Refuges, Big Game Refuges, Game Ranges and Wildlife Ranges administered by the U.S. Fish and Wildlife Service, and Wilderness and Primitive areas administered by the U.S. Forest Service. (PLT376) — AIM ¶7-4-6

Answers

2051 [A]	2052 [A]	2055 [B]	2056 [A]	2057 [A]	2058 [A]

ALL

2059. (Refer to Figure 60, point 6) The floor of the Class E airspace over the town of Commerce is

A—1,200 feet MSL.
B—700 feet AGL.
C—1,200 feet AGL.

The magenta shading around Commerce indicates the floor of the Class E airspace starts 700 feet above the surface. (PLT064) — Sectional Chart

Answers (A) and (C) are incorrect because 1,200 feet AGL (not MSL) is indicated with blue shading.

ALL

2114. (Refer to Figure 59, area 1.) Identify the airspace over Lowe Airport.

A—Class G airspace—surface up to but not including 18,000 feet MSL.
B—Class G airspace—surface up to but not including 700 feet MSL, Class E airspace—700 feet to 14,500 feet MSL.
C—Class G airspace—surface up to but not including 1,200 feet AGL, Class E airspace—1,200 feet AGL up to but not including 18,000 feet MSL.

The blue circle in the bottom left corner of Figure 59 indicates Lowe airport lies in Class G airspace from the surface to 1,200 feet AGL, and Class E airspace from 1,200 feet AGL to 18,000 feet MSL. The blue shading of the Victor airways (east of the Mississippi) indicates Class E airspace from 1,200 feet AGL. (PLT064) — Sectional Chart

ALL

2116. (Refer to Figure 56, area 1) The Nalf Fentress (NFE) airport is in what type of airspace?

A—Class C
B—Class E
C—Class G

The Nalf Fentress (NFE) airport is halfway between areas 1 and 2 in Figure 56. It is inside a magenta dashed line indicating Class E airspace starts at the surface. (PLT064) — Sectional Chart

Answer (A) is incorrect because Class C airspace is indicated with a solid magenta line. Answer (C) is incorrect because Class G airspace would be indicated with solid (shaded) magenta or blue lines.

ALL

2111. Which is true concerning the blue and magenta colors used to depict airports on Sectional Aeronautical Charts?

A—Airports with control towers underlying Class A, B, and C airspace are shown in blue, Class D and E airspace are magenta.
B—Airports with control towers underlying Class C, D, and E airspace are shown in magenta.
C—Airports with control towers underlying Class B, C, D, and E airspace are shown in blue.

Reference FAA Legend 1. Airports having control towers are shown in blue, all others in magenta. (PLT064) — Sectional Chart

Answer (A) is incorrect because Class A airspace is not depicted on a Sectional; Class C airspace is magenta; and Class E airspace is both blue and magenta depending on where the floor of the airspace starts. Answer (B) is incorrect because airports with control towers are shown in blue.

ALL

2198. Which is true concerning the colors used to depict airports on Sectional Aeronautical Charts?

A—Airports with control towers underlying Class D and E airspace are magenta.
B—Airports with control towers are shown in magenta.
C—Airports with control towers are shown in blue.

Reference FAA Legend 1 in the Computer Testing Supplement. Airports having control towers are shown in blue, all others in magenta. (PLT064) — Sectional Chart

ALL

2112. (Refer to Figure 56, area 2.) The flag symbol at Lake Drummond represents a

A—compulsory reporting point for Norfolk Class C airspace.
B—compulsory reporting point for Hampton Roads Airport.
C—visual checkpoint used to identify position for initial callup to Norfolk Approach Control.

Reference FAA Legend 1. The flag symbol represents a visual checkpoint used to identify position for initial callup to Norfolk Approach Control. (PLT064) — Sectional Chart

Answers

2059 [B]	2114 [C]	2116 [B]	2111 [C]	2198 [C]	2112 [C]

ALL

2113. (Refer to Figure 57, area 7.) The airspace overlying McKinney (TKI) is controlled from the surface to

A—700 feet AGL.
B—2,900 feet MSL.
C—2,500 feet MSL.

The airspace overlying McKinney (TKI) is controlled from the surface to 2900 feet MSL, as indicated by the [29] next to the airport information. (PLT064) — Sectional Chart

ALL

2115. (Refer to Figure 61, point 6) Mosier Airport is

A—an airport restricted to use by private and recreational pilots.
B—a restricted military stage field within restricted airspace.
C—a nonpublic use airport.

The "Pvt" after the airport names indicates Mosier airport is a restricted or non-public use airport. (PLT064) — Sectional Chart

Answer (A) is incorrect because the R in the circle indicates the airport is a nonpublic-use airport. Answer (B) is incorrect because the blue box near Mosier airport indicates an alert area.

ALL

2204. (Refer to Figure 61, point 1.) The floor of the Class E airspace above Georgetown Airport (Q61) is at

A—the surface.
B—700 feet AGL.
C—3,823 feet MSL.

Georgetown Airport is outside the magenta shaded area, which indicates the floor of Class E airspace is at 1,200 feet AGL. The airport elevation is given in the airport data as 2,623 feet MSL. Therefore, the Class E airspace above Georgetown Airport is 3,823 feet MSL (2,623 + 1,200). (PLT064) — Sectional Chart

Answer (A) is incorrect because Class E airspace begins at the surface only when surrounded by a magenta segmented circle. Answer (B) is incorrect because Class E airspace begins at 700 feet AGL inside the magenta shaded areas.

ALL

2138. (Refer to Figure 66, area 1.) The visibility and cloud clearance requirements for a Sport Pilot to operate over Sandpoint Airport at less than 700 feet AGL are

A—3 miles and clear of clouds.
B—1 mile and 1,000 feet above, 500 feet below, and 2,000 feet horizontally from each cloud.
C—3 miles and 1,000 feet above, 500 feet below, and 2,000 feet horizontally from each cloud.

The Sandpoint Airport Class E airspace starts at 700 feet AGL (indicated by the magenta shading). Class G airspace exists below this. The Sport Pilot clearance requirements are 3 miles visibility and clear of clouds. (PLT163) — 14 CFR §61.315

ALL

2179. Airports with control towers are shown in

A—magenta.
B—blue.
C—brown.

Reference FAA Legend 1. Airports having control towers are shown in blue, all others in magenta. (PLT064) — Sectional Chart

LTA

2294. A balloon flight through a restricted area is

A—permitted at certain times, but only with prior permission by the appropriate authority.
B—permitted anytime, but caution should be exercised because of high-speed military aircraft.
C—never permitted.

Restricted Areas denote the presence of unusual, often invisible, hazards to aircraft such as artillery firing, aerial gunnery, or guided missiles. Penetration of Restricted Areas without authorization of the using or controlling agency may be extremely hazardous to the aircraft and its occupants. (PLT393) — AIM ¶3-4-3

Answers

2113 [B]	2115 [C]	2204 [C]	2138 [A]	2179 [B]	2294 [A]

Chapter 6
Procedures and Operations

Preflight Procedures

Before beginning a flight, the pilot-in-command is required to become familiar with all available information concerning that flight. This information must include the following:

1. Runway lengths, and
2. Takeoff and landing information for airports of intended use, including aircraft performance data.

If the flight will not be in the vicinity of an airport, the pilot must also consider the following:

1. Weather reports and forecasts,
2. Fuel requirements (enough fuel to fly to the first point of intended landing, and at normal cruising speed, to fly after that for at least 30 minutes), and
3. Alternatives available if the flight cannot be completed as planned.

A thorough **routine preflight inspection** should be performed on an aircraft to ensure that the aircraft is prepared for safe flight. This should be a systematic means by which the pilot determines the airplane is ready for safe flight. Prior to every flight the same day, a pilot should perform at least a visual walk-around inspection of the aircraft.

After an aircraft has been stored for an extended period of time, a special check should be made during preflight for damage or obstructions caused by animals, birds, or insects.

For any aircraft with an unknown history, a more detailed annual or 100-hour inspection should be performed to evaluate all structural integrity and systems operation.

The use of a written checklist for preflight inspection and starting the engine is recommended to ensure that all necessary items are checked in a logical sequence.

Although 14 CFR Part 91 places primary responsibility on the owner or operator for maintaining an aircraft in an airworthy condition, the pilot-in-command is responsible for determining whether that aircraft is in condition for safe flight.

ALL

2132. Which preflight action is specifically required of the pilot prior to each flight?

A—Check the aircraft logbooks for appropriate entries.
B—Become familiar with all available information concerning the flight.
C—Review wake turbulence avoidance procedures.

Each PIC shall, before each flight, become familiar with all available information concerning that flight. This information must include:

(a) For a flight not in the vicinity of an airport, weather reports and forecasts, fuel requirements, alternatives available if the planned flight cannot be completed, and any known traffic delays of which the pilot has been advised by ATC;

(b) For any flight, runway lengths of airports of intended use, and the following takeoff and landing distance information:

1. For civil aircraft for which an approved airplane or rotorcraft flight manual containing takeoff and landing distance data is required, the takeoff and landing distance data contained therein; and

2. For civil aircraft other than those specified in paragraph (b)(1) of this regulation, other reliable information appropriate to the aircraft, relating to aircraft performance under expected values of airport elevation and runway slope, aircraft gross weight, and wind and temperature.

(PLT445) — 14 CFR §91.103

Answers
2132 [B]

ALL

2147. Preflight action, as required for all flights away from the vicinity of an airport, shall include

A—the designation of an alternate airport.
B—a study of arrival procedures at airports of intended use.
C—an alternate course of action if the flight cannot be completed as planned.

Each pilot-in-command shall, before each flight, become familiar with all available information concerning that flight. This information must include:

(a) For a flight not in the vicinity of an airport, weather reports and forecasts, fuel requirements, alternatives available if the planned flight cannot be completed, and any known traffic delays of which the pilot has been advised by ATC;

(b) For any flight, runway lengths of airports of intended use, and the following takeoff and landing distance information:

1. For civil aircraft for which an approved airplane or rotorcraft flight manual containing takeoff and landing distance data is required, the takeoff and landing distance data contained therein; and

2. For civil aircraft other than those specified in paragraph (b)(1), other reliable information appropriate to the aircraft, relating to aircraft performance under expected values of airport elevation and runway slope, aircraft gross weight, and wind and temperature.

(PLT445) — 14 CFR §91.103

ALL

2033. How should an aircraft preflight inspection be accomplished for the first flight of the day?

A—Quick walk around with a check of gas and oil.
B—Any sequence as determined by the pilot-in-command.
C—Thorough and systematic means recommended by the manufacturer.

The preflight inspection should be a thorough and systematic means by which the pilot determines that the airplane is ready for safe flight. Most Aircraft Flight Manuals or Pilot's Operating Handbooks contain a section devoted to a systematic method of performing a preflight inspection that should be used by the pilot for guidance. (H935) — FAA-H-8083-25

AIR, RTC, PPC, WSC

2220. Filling the fuel tanks after the last flight of the day is considered a good operating procedure because this will

A—force any existing water to the top of the tank away from the fuel lines to the engine.
B—prevent expansion of the fuel by eliminating airspace in the tanks.
C—prevent moisture condensation by eliminating airspace in the tanks.

Water in the fuel system is dangerous and the pilot must prevent contamination. The fuel tanks should be filled after each flight, or at least after the last flight of the day. This will prevent moisture condensation within the tank, since no air space will be left inside. (PLT445) — FAA-H-8083-25

Answer (A) is incorrect because water will settle to the bottom of a gas tank. Answer (B) is incorrect because fuel is allowed to expand by the fuel vent, whether or not the tanks are full.

Answers
2147 [C]   2033 [C]   2220 [C]

AIR, RTC, WSC, PPC

2338. To properly purge water from the fuel system of an aircraft equipped with fuel tank sumps and a fuel strainer quick drain, it is necessary to drain fuel from the

A—fuel strainer drain.
B—lowest point in the fuel system.
C—fuel strainer drain and the fuel tank sumps.

Many aircraft are equipped with fuel strainers located at the lowest point in the fuel lines and sump drains installed at the lowest point in each fuel tank. In order to completely purge all of the liquid water from the fuel system, the fuel strainer drain and the sumps in all of the tanks must be drained. (PLT253) — FAA-H-8083-25

AIR, RTC

2353. During preflight in cold weather, crankcase breather lines should receive special attention because they are susceptible to being clogged by

A—congealed oil from the crankcase.
B—moisture from the outside air which has frozen.
C—ice from crankcase vapors that have condensed and subsequently frozen.

The crankcase breather requires special consideration when preparing for cold weather. Frozen breather lines can create numerous problems. When crankcase vapors cool, they may condense in the breather line and subsequently freeze it closed. Special care is recommended during the preflight to ensure that the breather system is free of ice. (PLT126) — AC 91-13

Answer (A) is incorrect because oil that lays in the bottom of the crankcase never gets into the breather lines. Answer (B) is incorrect because a low air temperature is usually associated with a low moisture content.

AIR, RTC

2354. Which is true regarding preheating an aircraft during cold weather operations?

A—The cabin area as well as the engine should be preheated.
B—The cabin area should not be preheated with portable heaters.
C—Hot air should be blown directly at the engine through the air intakes.

Low temperatures may cause a change in the viscosity of engine oils, batteries may lose a high percentage of their effectiveness, and instruments may stick. Because of the above, preheating the engines as well as the cabin before starting is desirable in low temperatures. Extreme caution should be used in the preheat process to avoid fire. (PLT126) — AC 91-13

Answers
2338 [C] 2353 [C] 2354 [A]

Aircraft Lighting

Aircraft equipped with an **anti-collision strobe** light system are required to operate that light system during all types of operations (day and night). However, the pilot-in-command may determine that the anti-collision lights should be turned off when their operation would constitute a hazard to safety. The anticollision light may be either aviation red or aviation white. *See* Figure 6-1.

For collision avoidance, a pilot must know where each position colored light is located on an aircraft. For example, if a pilot observes a steady red light and a flashing red light ahead and at the same altitude, the other aircraft is crossing to the left; a steady white and a flashing red light indicates that the other aircraft is headed away from the observer; and steady red and green lights at the same altitude as the observer indicates that the other aircraft is approaching head-on. Many slower light-sport aircraft have blinking strobes used for anti-collision lighting for visual location. *See* Figure 6-2.

Figure 6-1. Example of anti-collision strobe lighting on LSA

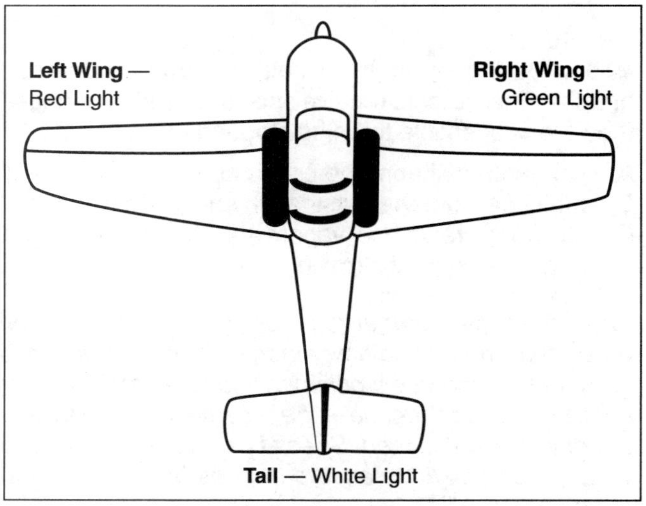

Figure 6-2. Airplane lighting

ALL

2020. Pilots must operate the anti-collision lights

A—at night or in inclement weather.

B—at night when the visibility is less than three miles and flying in Class B airspace.

C—day and night, except when the pilot-in-command determines that they constitute a hazard to safety.

Aircraft equipped with an anti-collision light system are required to operate that light system during all types of operations (day and night). However, the pilot-in-command may determine that the anti-collision lights should be turned off when their operation would constitute a hazard to safety. (PLT119) — 14 CFR §91.209

Answers

2020 [C]

Surface Operations

Taxiing to or from the runway generally presents no problems during calm or light wind conditions. However, when taxiing in moderate to strong wind conditions, the airplane's control surfaces or weight-shift control wing must be used to counteract the effects of wind. In airplanes equipped with a nose wheel (tricycle-gear), use the following taxi procedures:

1. The elevator should be in the neutral position when taxiing into a headwind.
2. The upwind aileron should be held in the up position when taxiing in a crosswind (or the upwind wing will tend to be lifted).
3. The elevator should be held in the down position and the upwind aileron down when taxiing with a quartering tailwind (the most critical condition for a nosewheel-type airplane).

When an airplane equipped with a tailwheel is taxied into a headwind, the elevator should be held in the up position to hold the tail down. In a quartering tailwind, both the upwind aileron and the elevator should be in the down position. *See* Figure 6-3.

AIR
2228. When taxiing an airplane with strong quartering tailwinds, which aileron position should be used?

A—Neutral.
B—Aileron down on the side from which the wind is blowing.
C—Aileron up on the side from which the wind is blowing.

Taxiing with a quartering tailwind provides the most hazardous conditions. In this case, the elevator should be in the down position and the aileron on the upwind side should also be in the down position to keep the wing from lifting. (PLT112) — FAA-H-8083-3

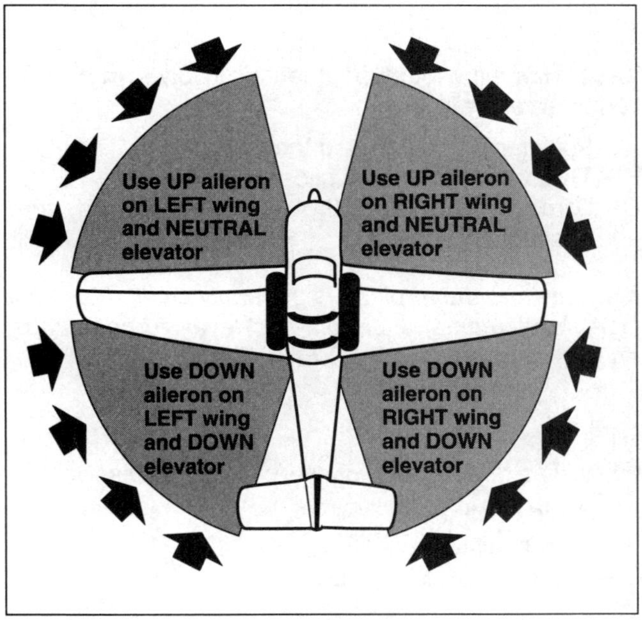

Figure 6-3. Control position while taxiing

Answers
2228 [B]

Aircraft Maneuvering

The four fundamentals involved in maneuvering an aircraft are: straight-and-level flight, turns, climbs, and descents. Climb depends upon the reserve power or thrust. Reserve power is the available power over and above that required to maintain horizontal flight at a given speed.

The most critical conditions for **takeoff performance** are the result of some combination of high gross weight, altitude, temperature, and unfavorable wind. In all cases, the pilot must make an accurate prediction of takeoff distance from the performance data of the AFM or POH, regardless of the runway available, and strive for a polished, professional takeoff procedure.

The **maximum endurance** condition is obtained at the point of minimum power required since this would require the lowest fuel flow to keep the aircraft in steady, level flight. Maximum range condition occurs where the proportion between speed and power required is greatest.

Total range is dependent on both fuel available and specific range. When range and economy of operation are the principal goals, the pilot must ensure that the aircraft is operated at the recommended long-range cruise condition. By this procedure, the aircraft will be capable of its maximum design-operating radius, or can achieve lesser flight distances with a maximum of fuel reserve at the destination.

ALL

2231. Name the four fundamentals involved in maneuvering an aircraft.

A—Power, pitch, bank, and trim.
B—Thrust, lift, turns, and glides.
C—Straight-and-level flight, turns, climbs, and descents.

Maneuvering the airplane is generally divided into four flight fundamentals: straight-and-level, turns, climbs, and descents. (PLT219) — FAA-H-8083-3

ALL

2233. The direct cause of every stall is excessive

A—angle of attack.
B—density altitude.
C—upward vertical velocity.

The direct cause of every stall is an excessive angle of attack. There are any number of flight maneuvers that may produce an increase in the angle of attack, but the stall does not occur until the angle of attack becomes excessive. (PLT477) — FAA-H-8083-25

ALL

2234. The most critical conditions of takeoff performance are the result of some combination of high gross weight, altitude, temperature, and

A—unfavorable wind.
B—obstacles surrounding the runway.
C—powerplant systems.

The most critical conditions for takeoff performance are the result of some combination of high gross weight, altitude, temperature, and unfavorable wind. In all cases, the pilot must make an accurate prediction of takeoff distance from the performance data of the AFM or POH, regardless of the runway available, and strive for a polished, professional takeoff procedure. (PLT134) — FAA-H-8083-25

Answers

2231 [C] 2233 [A] 2234 [A]

Communication Procedures

The key to operating at an airport without an operating control tower is selection of the correct **Common Traffic Advisory Frequency (CTAF)**. Pilots of inbound traffic should monitor and communicate as appropriate on the designated CTAF from 10 miles to landing. Pilots of departing aircraft should monitor/communicate on the appropriate frequency from start-up, during taxi, and until 10 miles from the airport unless the local procedures require otherwise. The CTAF is identified in appropriate aeronautical publications and on the sectional chart. If the airport has a part-time air traffic control tower (ATCT), the CTAF is usually a tower frequency. If a flight service station (FSS) is located on the airport, they will usually monitor that frequency and provide advisories when the tower is closed. Where there is no tower or FSS, UNICOM, (if available) is usually the CTAF. UNICOM is limited to the necessities of safe and expeditious operation of private aircraft pertaining to runways and wind conditions, types of fuel available, weather, and dispatching. Secondarily, communications may be transmitted concerning ground transportation, food, lodging and services available during transit. When no tower, FSS, or UNICOM is available, use MULTICOM frequency 122.9 for self-announce procedures. *See* Figure 6-4.

Flight Service Stations (FSS's) are air traffic facilities that provide pilot briefings, enroute communications and VFR search-and-rescue services, assist lost aircraft and aircraft in emergency situations, relay ATC clearances, originate Notices to Airmen, broadcast aviation weather and National Airspace System (NAS) information, and monitor navigational aids (NAVAIDs). FSS can be contacted by phone at 1-800-WX-BRIEF or typically by radio at 122.2. In addition, at selected locations FSS's provide En Route Flight Advisory Service (Flight Watch), take weather observations, issue airport advisories, and advise Customs and Immigration of transborder flights.

Automatic Terminal Information Service (ATIS) is a continuous broadcast of non-control information in selected high-activity terminal areas at larger tower-controlled airports. To relieve frequency congestion, pilots are urged to listen to ATIS, and on initial contact with ground control, to advise controllers that the information has been received by repeating the alphabetical code word appended to the broadcast. For example: "information Sierra received." The phrase "have numbers" does not indicate receipt of the ATIS broadcast and should not be used.

Continued

Facility at Airport	Frequency Use	Communication/Broadcast Procedures	
		Outbound	*Inbound*
1 UNICOM (No Tower or FSS)	Communicate with UNICOM station on published CTAF frequency (122.7, 122.8, 122.725, 122.975, or 123.0). If unable to contact UNICOM station, use self-announce procedures on CTAF.	Before taxiing and before taxiing on the runway for departure.	10 miles out. Entering downwind, base, and final. Leaving the runway.
2 No Tower, FSS, or UNICOM	Self-announce on MULTICOM frequency 122.9.	Before taxiing and before taxiing on the runway for departure.	10 miles out. Entering downwind, base, and final. Leaving the runway.
3 No Tower in operation, FSS open	Communicate with FSS on CTAF frequency.	Before taxiing and before taxiing on the runway for departure.	10 miles out. Entering downwind, base, and final. Leaving the runway.
4 FSS closed (No Tower)	Self-announce on CTAF.	Before taxiing and before taxiing on the runway for departure.	10 miles out. Entering downwind, base, and final. Leaving the runway.
5 Tower or FSS not in operation	Self-announce on CTAF.	Before taxiing and before taxiing on the runway for departure.	10 miles out. Entering downwind, base, and final. Leaving the runway.

Figure 6-4. Summary of recommended communications procedures

Enroute Flight Advisory Service (EFAS) provides aircraft with weather advisories pertinent to the route and altitude being flown. To obtain this service, pilots flying below FL180 should contact "flight watch" on frequency 122.0 MHz followed by the aircraft identification and the name of the nearest VOR to their position. A VOR is a radio navigation aid. Sport Pilots may obtain the location of a VOR facility from a GPS if necessary (by using the "Nearest" function) for the purpose of identifying the nearest VOR facility when calling Flight Watch. Example:

"Cleveland flight watch, Cessna one two three four Kilo, Mansfield VOR, over."

The basics steps for aircraft radio transmission are to state:

1. Who you are calling.
2. Who and what you are (can provide basic speed range for slower aircraft).
3. Where you are.
4. What your intentions are.

In Step 2, be more detailed about the aircraft you are flying. When flying the slower powered parachutes and weight-shift aircraft, you could include the pattern speed in your first transmission if you feel this would be in the interest of safety.

In following the steps above for your transmission, use these specific formats to communicate your information:

1. Name the facility being called. When calling a ground station, begin with the name of the facility followed by the type of facility being called. For example: When calling an FAA flight service station: "Seattle Radio"; or when calling a UNICOM: "Carson Area Traffic."
2. Civil aircraft pilots should state the aircraft type, model or manufacturer's name followed by the digits/letters of the registration number using the Phonetic Alphabet. For example: when flying Easy N2959S, the proper phraseology for initial contact with Buffalo Approach Control is: "Buffalo Approach, Easy two niner five niner Sierra." (The prefix "N" is dropped when the manufacturer's name or model is stated.) For slower light-sport aircraft, state the category (Weight Shift or Powered Parachute) and average pattern speed in knots: "Powered Parachute, aircraft ID, flying 30 knots."
3. When indicating where you are, state your position relative to the ATC you are calling: "12 statute miles northwest of Carson Airport, seven thousand feet."
4. When stating your intentions, simply say what you plan or want to do. At an airport with an ATC tower: "Want clearance to proceed into Class D airspace and land at Tahoe Airport" (at tower-controlled airports they tell you what and where to go and when to report to them). You must repeat what the controller told you to verify you understood correctly. When approaching a nontowered airport in Class E and G airspace, for example: "Intend to enter the left-hand pattern at forty-five degrees in ten minutes" or "Transitioning through North to South 4 miles west of airport at seven thousand feet."

When transmitting an altitude to ATC (up to but not including 18,000 feet MSL), state the separate digits of the thousands, plus the hundreds. For example, 4,500 is "four thousand five hundred."

An ATC clearance means an authorization by air traffic control for the purpose of preventing collision between known aircraft, for an aircraft to proceed under specified traffic conditions within controlled airspace.

At airports with operating air traffic control towers (ATCT), approval must be obtained prior to moving an aircraft onto the movement area. Ground control frequencies are provided to reduce congestion on the tower frequency. They are used for issuance of taxi information, clearances and other necessary contacts. If instructed by ground control to "taxi to" a particular runway, the pilot may proceed via taxiways and across runways to, but not onto, the assigned runway.

Aircraft arriving at an airport where a control tower is in operation should not change to ground control frequency until directed to do so by ATC.

In an emergency situation, pilots should squawk 7700 on their transponder (if so equipped) and contact 121.5.

ALL

2001. Inbound to an airport with no tower in operation but with a Flight Service Station (FSS) open, a pilot should communicate with the FSS on the common traffic advisory frequency (CTAF)

A—20 miles out.
B—10 miles out.
C—5 miles out.

Pilots of inbound traffic should monitor and communicate as appropriate on the designated CTAF from 10 miles to landing. Pilots of departing aircraft should monitor/communicate on the appropriate frequency from start-up, during taxi, and until 10 miles from the airport unless the local procedures require otherwise. (PLT170) — AIM ¶4-1-9

ALL

2205. What services will a FSS provide?

A—Fuel pricing.
B—Assistance during an emergency.
C—Clearance to taxi for takeoff.

Flight Service Stations (FSS's) are air traffic facilities that provide pilot briefings, enroute communications and VFR search-and-rescue services, assist lost aircraft and aircraft in emergency situations, relay ATC clearances, originate Notices to Airmen, broadcast aviation weather and National Airspace System (NAS) information, receive and process IFR flight plans, and monitor navigational aids (NAVAIDs). In addition, at selected locations FSS's provide En Route Flight Advisory Service (Flight Watch), take weather observations, issue airport advisories, and advise Customs and Immigration of transborder flights. (PLT172) — AIM ¶4-1-3

Answer (A) is incorrect because this is done by UNICOM. Answer (C) is incorrect because this is done by Ground Control or Tower Control.

Answers
2001 [B] 2205 [B]

ALL

2003. An ATC clearance means an authorization by ATC for an aircraft to proceed under specified conditions within

A—controlled airspace.
B—uncontrolled airspace.
C—published Visual Flight Rules (VFR) routes.

An ATC clearance means an authorization by air traffic control for the purpose of preventing collision between known aircraft, for an aircraft to proceed under specified traffic conditions within controlled airspace. (PLT044) — Pilot/Controller Glossary

Answer (B) is incorrect because ATC does not provide clearances in uncontrolled airspace. Answer (C) is incorrect because an ATC clearance is not required to fly these routes.

ALL

2004. If faced with an emergency where Air Traffic Control (ATC) assistance is desired and not already in contact, which frequency can be used to establish communications?

A—121.5 MHz.
B—122.5 MHz.
C—128.725 MHz.

Although the frequency in use or other frequencies assigned by ATC are preferable, the following emergency frequencies can be used for distress or urgency communications: 121.5 MHz, 243.0 MHz, or 2182 kHz. (PLT116) — AIM ¶6-3-1

ALL

2005. When a distress or urgency condition is encountered, the pilot of an aircraft with a coded radar beacon transponder, who desires to alert a ground radar facility, should squawk code

A—7700
B—7600
C—7500

When a distress or urgency condition is encountered, the pilot of an aircraft with a coded radar beacon transponder, who desires to alert a ground radar facility, should squawk Code 7700 and then immediately establish communications with the ATC facility. (PLT497) — AIM ¶6-2-2

Answer (B) is incorrect because 7600 is used for lost communications. Answer (C) is incorrect because 7500 is used to indicate a hijack situation.

ALL

2006. When outbound from an airport with a UNICOM station on the published common traffic advisory frequency (CTAF) and there is no tower or Flight Service Station (FSS), the pilot should contact UNICOM or use self-announce procedures on CTAF before

A—engine start.
B—taxiing and before taxiing on the runway.
C—the preflight inspection.

Pilots should contact UNICOM before taxiing, crossing any runway, and again before taxiing on the runway for departure. (PLT435) — AIM ¶4-1-9

ALL

2007. When outbound from an airport without a UNICOM station, tower or Flight Service Station (FSS), the pilot should self-announce on frequency

A—122.7
B—122.9
C—122.8

Where there is no tower, FSS, or UNICOM station on the airport, use MULTICOM frequency 122.9 for self-announce procedures. (PLT435) — AIM ¶4-1-9

Answers (A) and (C) are incorrect because these are both CTAF frequencies used to communicate with a UNICOM station.

ALL

2013. (Refer to Figure 66, area 2; and Figure 67.) At Coeur D'Alene, which frequency should be used as a Common Traffic Advisory Frequency (CTAF) to monitor airport traffic?

A—122.05 MHz.
B—135.075 MHz.
C—122.8 MHz.

CTAF is designed for the purpose of carrying out airport advisory practices and/or position reporting at an uncontrolled airport (which may also occur during hours when a tower is closed). The CTAF may be a UNICOM, MULTICOM, FSS, or tower frequency and is identified in the appropriate aeronautical publications. A solid dot with the letter "C" inside indicates Common Traffic Advisory Frequency. When the control tower operates part time and a UNICOM frequency is provided, use the UNICOM frequency. (PLT064) — AIM ¶4-1-9

Answer (A) is incorrect because 122.05 is the remote communication outlet (RCO) frequency to contact Boise FSS in the Coeur D'Alene vicinity. Answer (B) is incorrect because 135.075 is used to listen to the AWOS.

Answers

2003 [A]	2004 [A]	2005 [A]	2006 [B]	2007 [B]	2013 [C]

ALL

2014. (Refer to Figure 66, area 2; and Figure 67.) What is the correct UNICOM frequency to be used at Coeur D'Alene?

A—135.075 MHz.
B—122.05 MHz.
C—122.8 MHz.

The sectional chart and Airport Facility Directory both list the UNICOM frequency as 122.8. The UNICOM frequency is depicted in the airport information in the lower right-hand corner. (PLT064) — AIM ¶4-1-11

Answer (A) is incorrect because 135.075 is used to listen to AWOS. Answer (B) is incorrect because 122.05 is the remote communication outlet (RCO) frequency to contact Boise FSS in the Coeur D'Alene vicinity.

ALL

2015. You have just landed at a towered airport and the tower tells you to contact ground control when clear of the runway. You are considered clear of the runway when

A—all parts of the aircraft have crossed the hold line.
B—the aircraft cockpit is clear of the hold line.
C—the tail of the aircraft is clear of the runway edge.

An aircraft exiting a runway is not clear of the runway until all parts of the aircraft have crossed the applicable holding position marking. (PLT112) — AIM ¶2-3-5

ALL

2016. Pilots should state their position on the airport when calling the tower for takeoff

A—from a runway intersection, during instrument conditions.
B—from a runway intersection or the end of the runway.
C—from a runway intersection, only at night.

Pilots should state their position on the airport when calling the tower for takeoff from a runway intersection. ***Example:*** *Cleveland Tower, Apache Three Seven Two Two Papa, at the intersection of taxiway Oscar and runway two three right, ready for departure. (PLT486) — AIM ¶4-3-10*

Answers (A) and (C) are incorrect because this rule applies to all operations.

ALL

2017. After landing at a tower controlled airport a pilot should contact ground control

A—when advised by the tower.
B—prior to turning off the runway.
C—after reaching a taxiway that leads directly to the parking area.

A pilot who has just landed should not change from the tower frequency to the ground control frequency until directed to do so by the controller. (PLT170) — AIM ¶4-3-14

ALL

2034. (Refer to Figure 56, area 3.) What is the recommended communications procedure for a landing at Currituck County Airport?

A—Transmit intentions on 122.9 MHz when 10 miles out and give position reports in the traffic pattern.
B—Contact Elizabeth City FSS for airport advisory service.
C—Contact New Bern FSS for area traffic information.

Where there is no tower, FSS, or UNICOM station on the airport, use MULTICOM frequency 122.9 for self-announce procedures. Such airports will be identified with the circled C on the Sectional. (PLT064) — AIM ¶4-1-9

Answer (B) is incorrect because there is no Elizabeth City FSS (Elizabeth City uses the Raleigh FSS, indicated by "Raleigh" below the Elizabeth City VOR identifier box). Answer (C) is incorrect because the circled C indicates 122.9 is the CTAF to be used.

Answers

2014 [C]	2015 [A]	2016 [B]	2017 [A]	2034 [A]

Airport Traffic Area Communications and Light Signals

Unless otherwise authorized, aircraft are required to maintain two-way radio communication with the airport traffic control towers (ATCT) when operating to, from, or on the controlled airport, regardless of the weather. If radio contact cannot be maintained, ATC will direct traffic by means of light gun signals. *See* Figure 6-5.

If the aircraft radios fail while inbound to a tower-controlled airport, the pilot should remain outside or above the airport traffic area until the direction and flow of traffic has been determined and then join the airport traffic pattern and watch the tower for light signals.

The general warning signal (alternating red and green) may be followed by any other signal. For example, while on final approach for landing, an alternating red and green light followed by a flashing red light is received from the control tower. Under these circumstances, the pilot should abandon the approach, realizing the airport is unsafe for landing.

Color & Type of Signal	Movement of Vehicles, Equipment & Personnel	Aircraft on the Ground	Aircraft in Flight
Steady green	Cleared to cross, proceed or go	Cleared for takeoff	Cleared to land
Flashing green	Not applicable	Cleared for taxi	Return for landing (to be followed by steady green at the proper time)
Steady red	STOP	STOP	Give way to other aircraft and continue circling
Flashing red	Clear the taxiway/runway	Taxi clear of the runway in use	Airport unsafe, do not land
Flashing white	Return to starting point on airport	Return to starting point on airport	Not applicable
Alternating red and green	Exercise extreme caution	Exercise extreme caution	Exercise extreme caution

Figure 6-5

ALL

2012. A steady red light from the tower, for an aircraft on the ground indicates

A—Give way to other aircraft and continue circling.
B—Stop.
C—Taxi clear of the runway in use.

A steady red light from the tower, for an aircraft on the ground indicates the aircraft should stop. (PLT044) — 14 CFR §91.125

Answer (A) is incorrect because this would be issued to an aircraft in flight. Answer (C) is incorrect because this would be a flashing red light.

ALL

2352. A steady green light signal directed from the control tower to an aircraft in flight is a signal that the pilot

A—is cleared to land.
B—should give way to other aircraft and continue circling.
C—should return for landing.

A steady green light signal directed to an aircraft in flight means the pilot is cleared to land. (PLT502) — AIM ¶4-3-13

Answer (B) is incorrect because this would be a steady red signal. Answer (C) is incorrect because this would be a flashing green signal.

Answers
2012 [B]   2352 [A]

Traffic Advisories

Radar-equipped ATC facilities, typically in Class B and C airspace, provide traffic advisories and limited vectoring (called "basic radar service") to VFR aircraft, provided the aircraft can communicate with the facility, is within radar coverage, and can be radar-identified.

Traffic advisories given by a radar service will refer to the other aircraft by azimuth in terms of the 12-hour clock, with 12 o'clock being the direction of flight (track), not aircraft heading. Each hour is equal to 30°. For example, an aircraft heading 090° is advised of traffic at the 3 o'clock position. The pilot should look 90° to the right of the direction of flight, or to the south.

These radar programs are not to be interpreted as relieving pilots of their responsibilities to see and avoid other traffic operating in basic VFR weather conditions, to maintain appropriate terrain and obstruction clearance, or to remain in weather conditions equal to or better than the minimum required by the FAA regulations. Whenever compliance with an assigned route, heading, and/or altitude is likely to compromise pilot responsibilities respecting terrain, obstruction clearance, and weather minimums, the controller should be so advised and a revised clearance or instruction obtained.

ALL

2054. An ATC radar facility issues the following advisory to a pilot flying on a heading of 270°: 'TRAFFIC 3 O'CLOCK, 2 MILES, EASTBOUND...' Where should the pilot look for this traffic?

A—North.
B—South.
C—West.

Traffic information will be given in azimuth from the aircraft in terms of the 12-hour clock. Thus, each hour would constitute an angle of 30°. Picture a clock in your lap; 3 o'clock is to your right, 9 o'clock is to your left, 12 o'clock is straight ahead, and 6 o'clock is behind you. If an aircraft is proceeding on a heading of 270° (west), traffic located at the 3 o'clock position would be 90° right of the nose, north of the aircraft. (PLT194) — AIM ¶4-1-14

ALL

2226. During departure, when visual separation is employed by Air Traffic Control (ATC), traffic is no longer a factor when

A—the other aircraft turns away or is on a diverging course.
B—visual contact with the other aircraft is lost.
C—the other aircraft is passed.

Traffic is no longer a factor when during approach phase the other aircraft is in the landing phase of flight or executes a missed approach; and during departure or en route, when the other aircraft turns away or is on a diverging course. (PLT116) — AIM ¶4-4-14

Answers

2054 [A] 2226 [A]

Collision Avoidance

Part 91 has established right-of-way rules, minimum safe altitudes, and VFR cruising altitudes to enhance flight safety. The pilot can contribute to collision avoidance by being alert and scanning for other aircraft. This is particularly important in the vicinity of an airport. **Scanning** the sky for other aircraft is a key factor in collision avoidance. Pilots must develop an effective scanning technique which maximizes visual capabilities. Because the eyes focus only on a narrow viewing area, effective scanning is accomplished with a series of short, regularly spaced eye movements. Each movement should not exceed 10°, and each area should be observed for at least one second. At night, scan slowly to permit the use of off-center vision. Prior to starting any maneuver, a pilot should visually scan the entire area for collision avoidance. Any aircraft that appears to have no relative motion and stays in one scan quadrant is likely to be on a collision course. If a target shows neither lateral nor vertical motion, but increases in size, take evasive action. When climbing or descending VFR on an airway, execute gentle banks, right and left, to provide for visual scanning of the airspace.

Most skeletal structures are supported by guy wires which are very difficult to see in good weather and can be invisible at dusk or during periods of reduced visibility. These wires can extend about 1,500 feet horizontally from a structure; therefore, all skeletal structures should be avoided horizontally by at least 2,000 feet.

The FAA Near Mid-Air Collision Report indicates that 81% of the incidents occurred in clear skies and unrestricted visibility conditions. It is essential to remember that if another aircraft appears to have no relative motion, it is likely to be on a collision course with you. If the other aircraft shows no lateral or vertical motion, but increases in size, take evasive action.

ALL
2065. Most midair collision accidents occur during

A—clear days.
B—hazy days.
C—cloudy nights.

The FAA Near Mid-Air Collision Report indicates that 81% of the incidents occurred in clear skies and unrestricted visibility conditions. (PLT194) — AC 90-48

ALL
2088. How can you determine if another aircraft is on a collision course with your aircraft?

A—The other aircraft will always appear to get larger and closer at a rapid rate.
B—The nose of each aircraft is pointed at the same point in space.
C—There will be no apparent relative motion between your aircraft and the other aircraft.

It is essential to remember that if another aircraft appears to have no relative motion, it is likely to be on a collision course with you. If the other aircraft shows no lateral or vertical motion, but increases in size, take evasive action. (PLT194) — AC 90-48

ALL
2066. Pilots who become apprehensive for their safety for any reason should

A—request assistance immediately.
B—reduce their situational awareness.
C—change their mindset.

Pilots who become apprehensive for their safety for any reason should request assistance immediately. Ready and willing help is available in the form of radio, radar, direction finding stations and other aircraft. Delay has caused accidents and cost lives. Safety is not a luxury! Take action! (PLT116) — AIM ¶6-1-2

Answers
2065 [A] 2088 [C] 2066 [A]

ALL

2184. Prior to starting each maneuver, pilots should

A—check altitude, airspeed, and heading.
B—visually scan the entire area for collision avoidance.
C—announce their intentions on the nearest CTAF.

Scanning the sky for other aircraft is a key factor in collision avoidance. (PLT194) — FAA-H-8083-25

Answer (A) is incorrect because checking your instruments is important, but secondary to collision avoidance. Answer (C) is incorrect because announcing your intentions to the nearest CTAF does not guarantee that anyone is listening.

ALL

2064. To scan properly for traffic, a pilot should

A—slowly sweep the field of vision from one side to the other at intervals.
B—concentrate on any peripheral movement detected.
C—use a series of short, regularly spaced eye movements that bring successive areas of the sky into the central visual field.

Effective scanning is accomplished with a series of short, regularly spaced eye movements that bring successive areas of the sky into the central visual field. Each movement should not exceed 10°, and each area should be observed for at least one second to enable detection. (PLT194) — AIM ¶8-1-6

ALL

2096. The most effective method of scanning for other aircraft for collision avoidance during daylight hours is to use

A—regularly spaced concentration on the 3-, 9-, and 12-o'clock positions.
B—a series of short, regularly spaced eye movements to search each 10-degree sector.
C—peripheral vision by scanning small sectors and utilizing offcenter viewing.

Effective scanning is accomplished with a series of short, regularly spaced eye movements that bring successive areas of the sky into the central visual field. Each movement should not exceed 10°, and each area should be observed for at least one second to enable detection. (PLT099) — AIM ¶8-1-6

ALL

2062. Guy wires, which support antenna towers, can extend horizontally; therefore, the towers should be avoided horizontally by at least

A—2,000 feet horizontally.
B—300 feet horizontally.
C—1,000 feet horizontally.

Most skeletal structures are supported by guy wires which are very difficult to see in good weather and can be invisible at dusk or during periods of reduced visibility. These wires can extend about 1,500 feet horizontally from a structure; therefore, all skeletal structures should be avoided horizontally by at least 2,000 feet. (PLT116) — AIM ¶7-5-3

Answers

2184 [B]	2064 [C]	2096 [B]	2062 [A]

Wake Turbulence

All aircraft leave two types of **wake turbulence**: Prop or jet blast, and wing-tip vortices.

The **prop** or **jet blast** could be hazardous to light aircraft on the ground behind large aircraft which are either taxiing or running-up their engines. In the air, jet or prop blast dissipates rapidly.

Wing-tip vortices are a by-product of lift. When a wing is flown at a positive angle of attack, a pressure differential is created between the upper and lower wing surfaces, and the pressure above the wing will be lower than the pressure below the wing. In attempting to equalize the pressure, air moves outward, upward, and around the wing tip, setting up a **vortex** which trails behind each wing. *See* Figure 6-6.

The strength of a vortex is governed by the weight, speed, and the shape of the wing of the generating aircraft. Maximum vortex strength occurs when the generating aircraft is heavy, clean, and slow.

Vortices generated by large aircraft in flight tend to sink below the flight path of the generating aircraft. A pilot should fly at or above the larger aircraft's flight path in order to avoid the wake turbulence created by the wing-tip vortices. *See* Figure 6-7.

Close to the ground, vortices tend to move laterally. A crosswind will tend to hold the upwind vortex over the landing runway, while a tailwind may move the vortices of a preceding aircraft forward into the touchdown zone.

To avoid wake turbulence when landing, a pilot should note the point where a preceding large aircraft touched down and then land past that point. *See* Figure 6-8.

On takeoff, lift off should be accomplished prior to reaching the rotation point of a preceding large aircraft; the flight path should then remain upwind and above the preceding aircraft's flight path. *See* Figure 6-9.

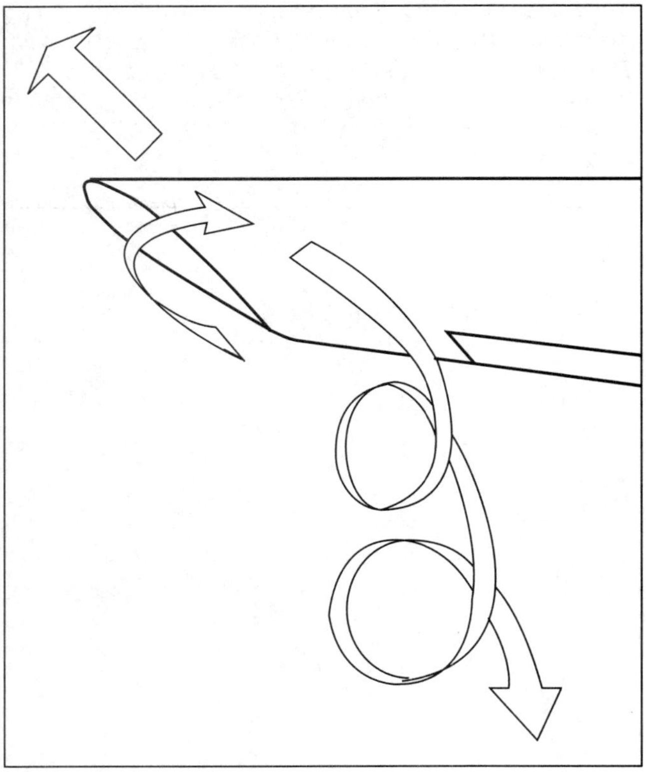

Figure 6-6. Wing-top vortices

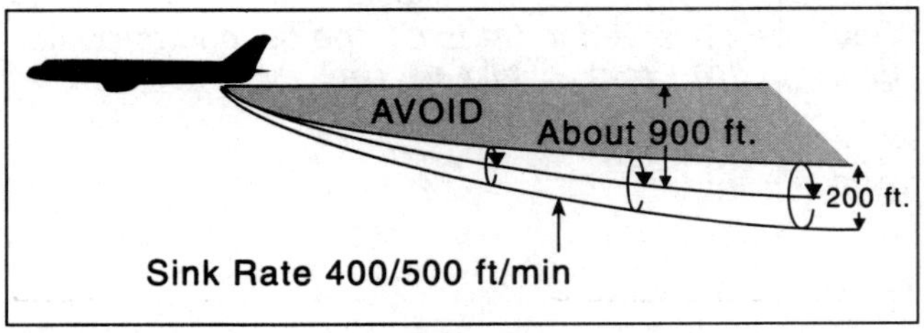

Figure 6-7. Vortices in cruise flight

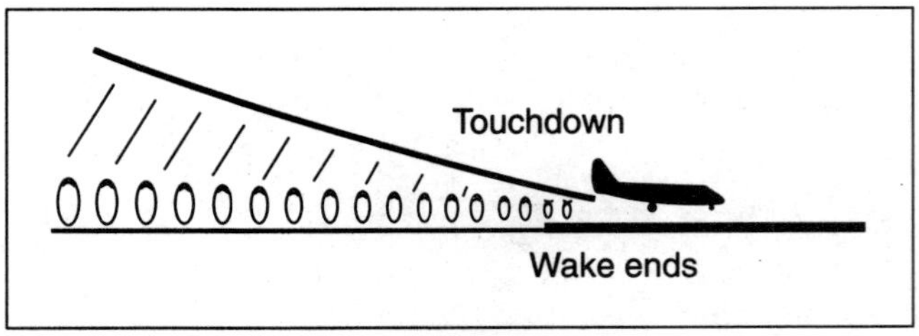

Figure 6-8. Touchdown and wake end

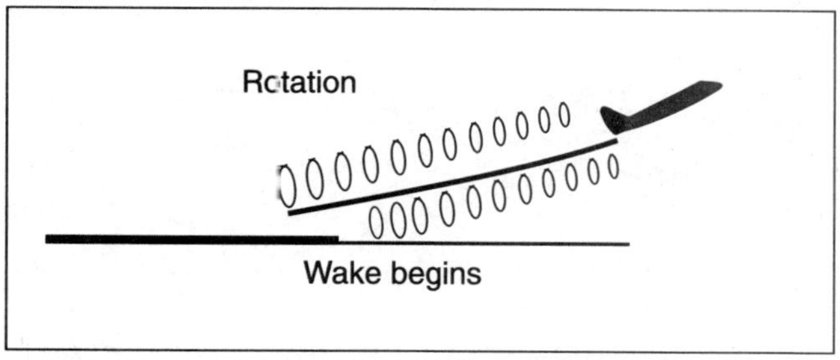

Figure 6-9. Rotation and wake beginning

ALL

2039. What wind condition prolongs the hazards of wake turbulence on a landing runway for the longest period of time?

A—Light quartering headwind.
B—Direct tailwind.
C—Light quartering tailwind.

A light wind with a cross runway component of 1 to 5 knots could result in the upwind vortex remaining in the touchdown zone for a period of time and hasten the drift of the downwind vortex toward another runway. Similarly, a tailwind condition can move the vortices of the preceding aircraft forward into the touchdown zone. The light quartering tailwind requires maximum caution. (PLT509) — AIM ¶7-3-4

ALL

2070. When landing behind a large aircraft, the pilot should avoid wake turbulence by staying

A—above the large aircraft's final approach path and landing beyond the large aircraft's touchdown point.
B—below the large aircraft's final approach path and landing before the large aircraft's touchdown point.
C—above the large aircraft's final approach path and landing before the large aircraft's touchdown point.

When landing behind a large aircraft stay at or above the large aircraft's final approach path. Note its touchdown point and land beyond it. (PLT509) — FAA-H-8083-25

Answer (B) is incorrect because below the flight path, you will fly into the sinking vortices generated by the large aircraft. Answer (C) is incorrect because by landing before the large aircraft's touchdown point, you will have to fly below the preceding aircraft's flight path, and into the vortices.

ALL

2071. Wingtip vortices created by large aircraft tend to

A—sink below the aircraft generating turbulence.
B—rise into the traffic pattern.
C—rise into the takeoff or landing path of a crossing runway.

Flight tests have shown that the vortices from large aircraft sink at a rate of about 400 to 500 feet per minute. They tend to level off at a distance about 900 feet below the path of the generating aircraft. (PLT509) — FAA-H-8083-25

Answers (B) and (C) are incorrect because wing-tip vortices sink toward the ground; however, they may move horizontally depending on crosswind conditions.

ALL

2072. Wingtip vortices are created only when an aircraft is

A—operating at high airspeeds.
B—heavily loaded.
C—developing lift.

Lift is generated by the creation of a pressure differential over the wing surface. The lowest pressure occurs over the wing surface and the highest pressure occurs under the wing. This pressure differential triggers the roll up of the airflow aft of the wing, resulting in wing-tip vortices. Vortices are generated from the moment an aircraft leaves the ground, since trailing vortices are a by-product of wing lift. (PLT509) — FAA-H-8083-25

Answer (A) is incorrect because the greatest turbulence is produced from an airplane developing lift at a slow airspeed. Answer (B) is incorrect because even though a heavily loaded airplane may produce greater turbulence, an airplane does not have to be heavily loaded in order to produce wing-tip vortices.

ALL

2235. The greatest vortex strength occurs when the generating aircraft is

A—light, dirty, and fast.
B—heavy, dirty, and fast.
C—heavy, clean, and slow.

The strength of a vortex is governed by the weight, speed, and the shape of the wing of the generating aircraft. Maximum vortex strength occurs when the generating aircraft is heavy, clean, and slow. (PLT509) — FAA-H-8083-25

Answers

2039 [C]	2070 [A]	2071 [A]	2072 [C]	2235 [C]

Chapter 7
Cross-Country

Navigation

Air navigation is the art of directing an aircraft along a desired course and being able to determine its geographical position at any time. Such navigation may be accomplished by pilotage, dead reckoning, or using radio navigational aids.

The aeronautical charts most commonly used by VFR pilots are the **VFR Sectional Aeronautical Chart**, the **VFR Terminal Area Chart**, and the **World Aeronautical Chart**. All three charts include aeronautical information such as airports, airways, special use airspace, and other pertinent data.

The scale of the VFR Sectional Aeronautical Chart is 1:500,000 (1 inch = 6.86 NM). Designed for visual navigation of slow speed aircraft in VFR conditions, this chart portrays terrain relief and checkpoints such as populated places, roads, railroads, and other distinctive landmarks. These charts have the best detail and are revised every 6 months.

To identify a point on the surface of the earth, a geographic coordinate, or "grid" system was devised. By reference to meridians of **longitude** and parallels of **latitude**, any position may be accurately located when using the grid system. When planning a flight, measure the true course from the mid-meridian to get an average true course.

Because it correlates with the equal divisions of longitude and latitude, the nautical mile is the measurement most often used in general aviation. A nautical mile (NM) is about 15% longer than a statute mile, or SM (the measure commonly used in the U.S.): 100 NM × 1.15 = 115 SM. Nautical miles used in speed calculations result in knots. 50 knots (nautical miles per hour) is 50 × 1.15 = 57.5 miles per hour (MPH); 100 SM would be 100 ÷ 1.15 = 87 knots. 50 MPH ÷ 1.15 = 43.5 knots. Either system can be used, but the slower-speed light-sport aircraft use SM and MPH, while faster airplanes use knots and nautical miles. *See* Figure 71.

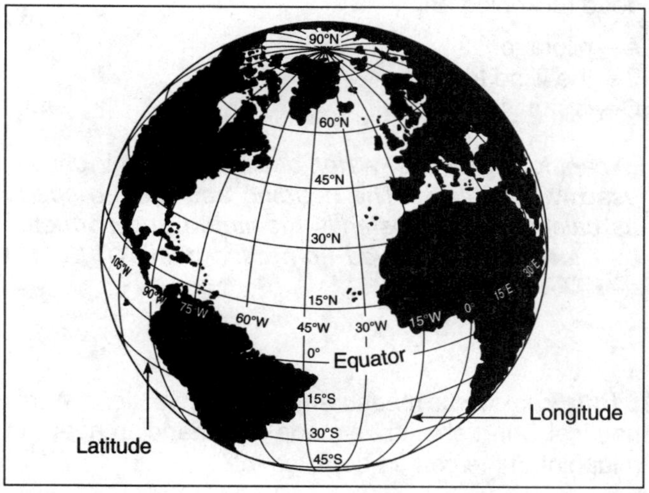

Figure 7-1. Meridians of longitude and parallels of latitude

GPS

Most of today's flight navigation systems installed in aircraft use the **Global Positioning System (GPS)**. This is becoming the standard from light-sport aircraft all the way to commercial jets. While flying, GPS provides your ground speed and ground track directly. Additional features of GPS are: time to location flying in the current wind, distance to location, and a direct path to the intended location (accounting for the wind). This is a valuable tool that aids basic visual reference, using the pilotage techniques described below.

Pilotage

Pilotage is the use of visible landmarks to maintain a desired course, and is the basic form of navigation for the pilot operating under Visual Flight Rules (VFR). Visible landmarks which can be identified on aeronautical charts allow the pilot to proceed from one checkpoint to the next.

ALL

2106. During VFR navigation without radio instruments, heading and groundspeed, as calculated by dead reckoning, should be constantly monitored and corrected by

A—pilotage as observed from checkpoints.
B—the wind triangle.
C—wet compass and the groundspeed indicator.

Except for flights over water, dead reckoning is usually used with pilotage. The heading and ground speed as calculated is constantly monitored and corrected by pilotage as observed from checkpoints. (PLT116) — FAA-H-8083-25

ALL

2108. For cross-country flights over land, navigation without radio instruments is usually accomplished using dead reckoning and

A—pilotage.
B—the wind triangle.
C—compass heading.

Except for flights over water, dead reckoning is usually used with pilotage. The heading and ground speed as calculated is constantly monitored and corrected by pilotage as observed from checkpoints. (PLT116) — FAA-H-8083-25

ALL

2100. True course measurements on a Sectional Aeronautical Chart should be made at a meridian near the midpoint of the course because the

A—values of isogonic lines change from point to point.
B—angles formed by isogonic lines and lines of latitude vary from point to point.
C—angles formed by lines of longitude and the course line vary from point to point.

Sectional Aeronautical Charts are made on grid in which the meridians of longitude are nonparallel straight lines that all meet at the earth's poles. When planning a flight, measure the true course from the mid-meridian to get an average true course. (PLT200) — FAA-H-8083-25

ALL

2178. Using pilotage, a pilot can keep from drifting too far off course by

A—flying the flight plan heading.
B—filing a flight plan.
C—referring to selected features or checkpoints.

Pilotage is navigation by reference to landmarks or checkpoints. The checkpoints selected should be prominent features common to the area of the flight and be readily identified by other features such as roads, rivers, and power lines. Select features that will make useful boundaries or brackets on each side of the course, such as highways, rivers, railroads, and mountains. (PLT335) — FAA-H-8083-25

ALL

2319. (Refer to Figure 58, area 2.) Which airport is located at approximately 47° 39 minutes 30 seconds N latitude and 100° 53 minutes 00 seconds W longitude?

A—Linrud.
B—Crooked Lake.
C—Johnson.

Graticules on Sectional Aeronautical Charts are the lines dividing each 30 minutes of latitude and each 30 minutes of longitude. Each tick mark represents one minute of latitude or longitude. Latitude increases northward, west longitude increases going westward. The Crooked Lake airport is located at approximately 47°39'30"N latitude and 100°53'00"W longitude. (PLT064) — FAA-H-8083-25

ALL

2306. (Refer to Figure 57, area 8.) What minimum altitude is required to fly over the Cedar Hill TV towers in the congested area south of NAS Dallas?

A—2,555 feet MSL.
B—3,449 feet MSL.
C—3,349 feet MSL.

The height of the tower is 2,449 MSL. A 1,000-foot clearance is required between obstructions and aircraft over a congested area. 2,449 MSL + 1,000 = 3,449 MSL (PLT064) — Sectional Chart Legend

Answers

2106 [A]	2108 [A]	2100 [C]	2178 [C]	2319 [B]	2306 [B]

LTA

2302. (Refer to Figure 56, area 4.) A balloon launched at the town of Edenton drifts northeasterly along the railroad. What minimum altitude must it maintain to clear all of the obstacles in the vicinity of Hertford by at least 500 feet?

A—805 feet MSL.
B—1,000 feet MSL.
C—1,015 feet MSL.

Use the following steps:

1. *Locate the towns of Edenton and Hertford.*
2. *Note the highest obstruction in the vicinity of Hertford (515 feet MSL).*
3. *Calculate the minimum altitude (MSL) by adding the required 500-foot clearance to the obstacle height.*

   515 feet + 500 feet = 1,015 feet MSL.

(PLT064) — AIM ¶7-5-3

LTA

2303. (Refer to Figure 58, area 1.) A balloon launched at Flying S Airport drifts southward towards the lighted obstacle. If the altimeter was set to the current altimeter setting upon launch, what should it indicate if the balloon is to clear the obstacle at 500 feet above the top?

A—1,531 feet MSL.
B—1,809 feet MSL.
C—3,649 feet MSL.

The height of the lighted obstacle is 3,149 feet MSL. The altimeter will indicate 3,649 (3149 + 500 if set to the current altimeter setting upon launch). (PLT041) — Sectional Chart Legend

LTA

2318. (Refer to Figure 58.) An airship crosses over Minot VORTAC (area 1) at 1056 and over the creek 8 nautical miles south-southeast on Victor 15 at 1108. What should be the approximate position on Victor 15 at 1211?

A—Over Lake Nettie National Wildlife Refuge.
B—Crossing the road east of Underwood.
C—Over the powerlines east of Washburn Airport.

This is a time/speed/distance problem.

1. *The distance is 8 NM (given in the question).*
2. *The time is: 11:08 – 10:56 = 12 minutes*
3. *Use: Distance ÷ Time x 60 = Ground speed*

   8 NM ÷ 12 min x 60 = 40 knots
4. *Calculate the distance traveled on the next leg at this speed by 12:11.*

   12:11 – 11:08 = :63 minutes
5. *Use: Speed ÷ 60 x Time = Distance*

   40 kts ÷ 60 min. x 63 min. = 42 NM
6. *Locate the 8 NM creek and measure an additional 42 NM along V15. You are east of Underwood; it is west of your position. The legs were 8 NM and 42 NM, total 50 NM.*

(PLT064) — FAA-H-8083-25

Answers

2302 [C] 2303 [C] 2318 [B]

LTA

2320. (Refer to Figure 60.) An airship passes over the Quitman VOR-DME area 2) at 0940 and then over the intersection of the powerline and Victor 114 at 0948. Approximately what time should the flight arrive over the Bonham VORTAC (area 3)?

A—1109.
B—1117.
C—1138.

Use the following steps:

1. *Locate Quitman VOR/DME (area 2) and Bonham VORTAC (area 3)*
2. *Measure the distance from Quitman VOR-DME to the intersection of the powerline with V114 (4 NM).*
3. *Calculate the ground speed of the airship:*

   Speed = Distance ÷ Time

   Speed = 4 NM ÷ (0948 – 0940)

   Speed = 4 NM ÷ 8 minutes

   Speed = 30 knots
4. *Measure the distance remaining from the "intersection" to Bonham VORTAC (55 NM).*
5. *Calculate the time required to travel 55 NM at a constant 30 knot ground speed.*

   Time = Distance ÷ Speed

   Time = 55 NM ÷ 30 knots

   Time = 1 hour 50 minutes
6. *Compute the arrival time by the adding estimated enroute time to the time of passing the intersection:*

   0150 time en route + 0948 time passing intersection = 1138 ETA.

(PLT012) — FAA-H-8083-25

LTA

2346. (Refer to Figure 66, area 2.) If a balloon is launched at Ranch Aero (Pvt) Airport with a reported wind from 220° at 5 knots, what should be its approximate position after 2 hours of flight?

A—Near Hackney (Pvt) Airport.
B—Crossing the railroad southwest of Granite Airport.
C—3-1/2 miles southwest of Rathdrum.

A balloon moves with the wind, so the problem asks: where do you end up if your ground speed is 5 knots and you are flying a true course of 220° for two hours?

1. *Calculate the distance that will be traveled at this speed during the time given.*

   Speed x Time = Distance

   5 NM x 2 hours = 10 NM total distance
2. *Draw your true course line from Ranch Aero, from 220°, and measure 10 NM up the line from Ranch Aero to find you are near Hackney (Pvt) Airport.*

(PLT012) — FAA-H-8083-25

Answers
2320 [C]   2346 [A]

Dead Reckoning

Dead reckoning is the method used for determining position with a heading indicator and calculations based on speed, elapsed time, and wind effect from a known position. The instruments used for dead reckoning navigation include the outside air temperature gauge, the airspeed indicator, the altimeter, the clock, and the magnetic compass system or slaved gyro system. These instruments provide information concerning direction, airspeed, altitude, and time, and must be correctly interpreted for successful navigation.

During flight, GPS provides a true course to the destination (accounting for wind). You follow this true course and it provides modern inflight dead reckoning. But for detailed flight planning, the following procedure should be used for plotting courses.

Plotting Courses

A course is the direction of flight measured in degrees clockwise from north. Meridians of longitude run from the south pole to the north pole. This alignment is called true north. When a course is plotted on a chart in relation to the lines of longitude and/or latitude it is called a **true course (TC)**, and will be expressed in three digits. North may be either 360° or 000°; east is 090°; south is 180°; and west is 270°. Any attempt to project lines of latitude and longitude onto a flat surface such as a chart results in a certain amount of distortion. When plotting a course on a sectional aeronautical chart, this distortion may be minimized by measuring true course in reference to the meridian nearest to the halfway point between the departure point and the destination.

A common type of plotter that is used to plot a course is a semicircular protractor with a straight edge attached to it. *See* Figure 7-2. The straight edge has distance scales that match various charts and these scales may depict both statute and nautical miles. A small hole at the base of the protractor portion indicates the center of the arc of the angular scale. Two complete scales cover the outer edge of the protractor, and they are graduated in degrees. An inner scale measures the angle from the vertical.

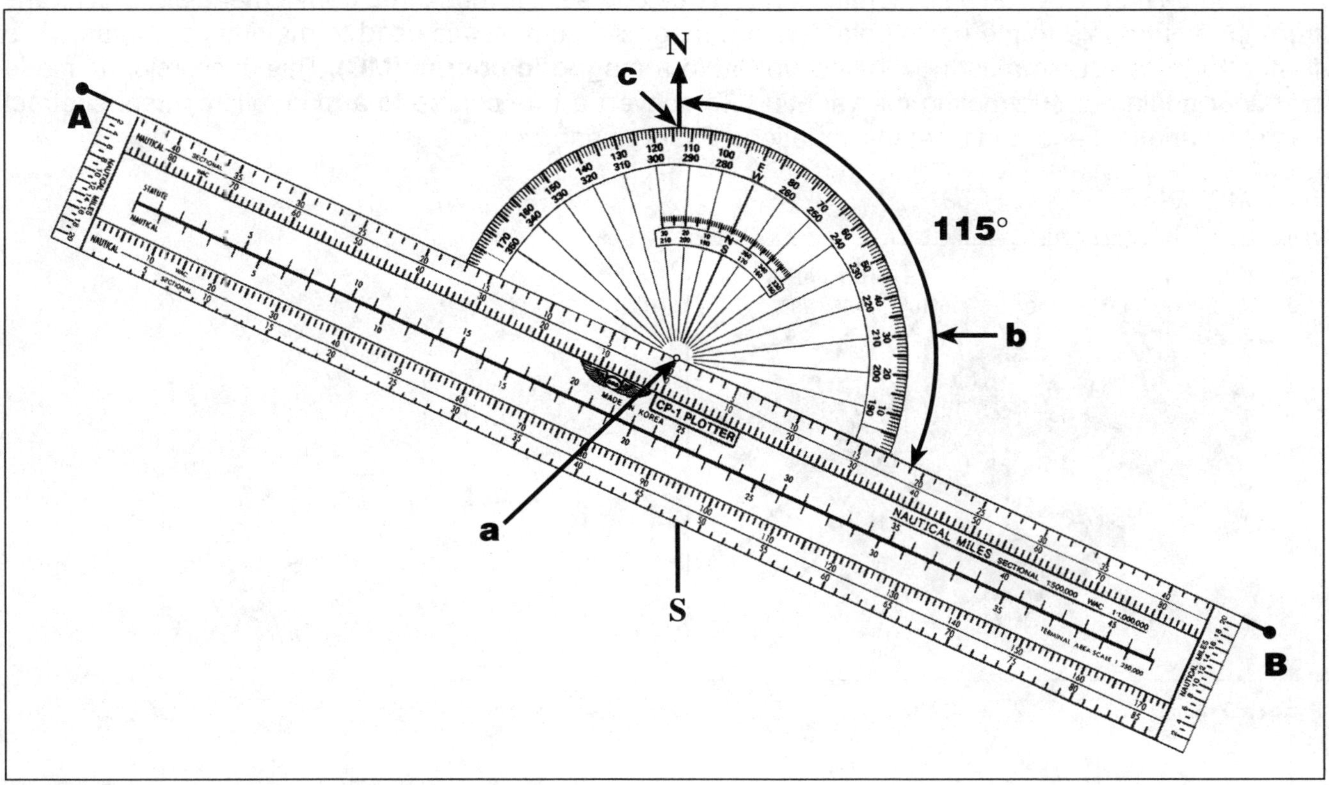

Figure 7-2. Placing the plotter on the course line

To determine true course (TC), use the plotter in the following manner:

1. Using the straight edge of the plotter as a guide, draw a line from the point of departure to the destination.
2. Place the top straight edge of the plotter parallel to the plotted course and move the plotter along the course line as necessary to place the small center hole over a meridian as near to the halfway point of the course as possible.
3. The true course is the angle measured between the meridian and the course line. The outer scale is used to read all angles between north through east to south, and the inner scale is used to read all angles between south through west to north.
4. Course lines which fall within 30° of true north or south can be more easily measured by sliding the plotter along the course line until the hole falls over a horizontal latitude line; use the partial protractor scale for these measurements.

Magnetic Variation

The north pole where all meridians converge is **true north**. The north pole which attracts the needle of a compass is **magnetic north**. These two poles are not in the same place. At any point where magnetic north and true north are in line with each other, the compass needle points both to magnetic north, and coincidentally, true north. The line along which this occurs is known as the agonic line. When positioned west of the agonic line, a compass will point right (east) of true north. When positioned east of the **agonic line**, a compass will point left (west) of true north. This angular difference between true north and magnetic north is called **magnetic variation (VAR)**. West of the agonic line, variation is "easterly." East of the agonic line, variation is "westerly." *See* Figure 7-3.

The amount and direction of variation is depicted on sectional charts as dashed magenta colored lines connecting points of equal variation, called isogonic lines.

A course measured on a sectional chart is a true course: it is measured from a meridian, which runs from the south pole to the north pole. Since a magnetic compass is used to maintain a course while flying, this true course must now be converted to a **magnetic course (MC)**. This conversion is made by either adding or subtracting the variation. To convert a true course to a magnetic course, subtract easterly variation, and add westerly variation: "East is least, west is best."

TC ± VAR = MC

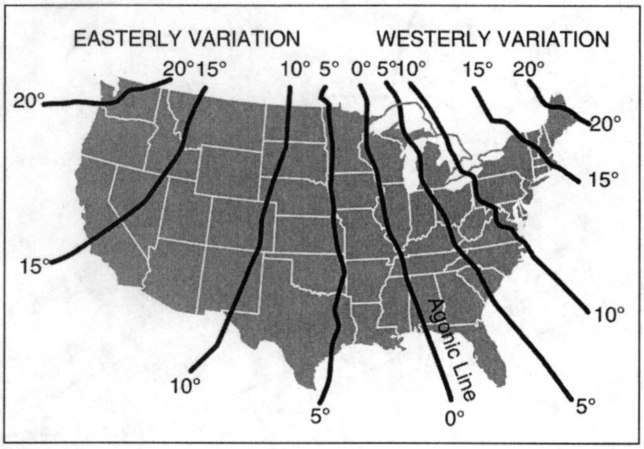

Figure 7-3

Magnetic Deviation

The magnetic compass is affected by influences within the aircraft such as electrical circuits, radios, engines, magnetized metal parts, etc., which cause the compass needle to be deflected from its normal reading. This deflection is known as **deviation (DEV)**, and it must be applied to convert a magnetic course to a **compass course (CC)** to make it usable in flight.

Deviation, which is different for each aircraft, may also vary for different courses in the same airplane. To let the pilot know the appropriate correction, a correction card is mounted near the compass.

To determine the actual compass reading to be followed during flight, it is necessary to apply the corrections for both variation and deviation:

True Course ± Variation = Magnetic Course ± Deviation = Compass Course

or,

TC ± VAR = MC ± DEV = CC

ALL

2201. Heading and groundspeed are calculated using dead reckoning procedures. In flight, they are constantly monitored and corrected by

A—pilotage as observed from checkpoints.
B—the wind triangle.
C—the wet compass and the groundspeed indicator.

Dead reckoning is navigation solely by means of computations based on time, airspeed, distance, and direction. The products derived from these variables, when adjusted by windspeed and velocity, are heading and ground speed. Except for flights over water, dead reckoning is usually used with pilotage for cross-country flying. The heading and ground speed as calculated is constantly monitored and corrected by pilotage as observed from checkpoints. (PLT335) — FAA-H-8083-25

ALL

2221. When converting from true course to magnetic heading, a pilot should

A—subtract easterly variation and right wind correction angle.
B—add westerly variation and subtract left wind correction angle.
C—subtract westerly variation and add right wind correction angle.

When converting a true course to a true heading, subtract a left wind correction angle or add a right wind correction angle. When converting from a true heading to a magnetic heading, add westerly variation or subtract easterly variation. (PLT012) — FAA-H-8083-25

Answer (A) is incorrect because right wind correction is added. Answer (C) is incorrect because westerly variation is added.

ALL

2102. The course measured on a sectional chart by reference to a meridian is known as the

A—true course.
B—magnetic course.
C—true heading.

The course measured on the chart is known as the true course. (PLT320) — FAA-H-8083-25

Answer (B) is incorrect because magnetic course adds variation to the course measured on the chart. Answer (C) is incorrect because true heading accounts for wind.

Answers

2201 [A] 2221 [B] 2102 [A]

Wind and Its Effects

An additional computation, common to both pilotage and dead reckoning, is necessary to compensate for the effect of wind.

Wind direction is reported as the direction from which the wind blows, i.e., wind blowing from the west to the east is a west wind. Wind speed is the rate of motion without regard to direction. In the United States, wind speed is usually expressed in knots. Wind velocity includes both direction and speed of the wind; for example, a west wind of 25 knots.

Downwind movement is with the wind; upwind movement is against the wind. Moving air exerts a force in the direction of its motion on any object within it. Objects that are free to move in air will move in a downwind direction at the speed of the wind.

If a powered aircraft is flying in a 20-knot wind, it will move 20 nautical miles downwind in 1 hour in addition to its forward movement through the air. The path of an aircraft over the earth is determined both by the motion of the aircraft through the air and by the motion of the air over the earth's surface. Direction and movement through the air is governed both by the direction that the aircraft nose is pointing and by aircraft speed.

The sideward displacement of an aircraft caused by wind is called drift. Drift can be determined by measuring the angle between the heading (the direction in which the nose is pointing) and the track (the actual path of the aircraft over the earth). *See* Figure 7-4.

For example, Figure 7-4 shows an aircraft which departs point A on a heading of 360° and flies for 1 hour in a wind of 270° at 20 knots. Under a no-wind condition the aircraft would arrive at point B at the end of 1 hour. However, this example has a wind of 20 knots, and the aircraft moves with the wind; so at the end of 1 hour, the aircraft is at point C, 20 nautical miles downwind from B. From A to B is the intended path of the aircraft, from B to C is the motion of the body of air, and from A to C is the actual path of the aircraft over the earth (the track).

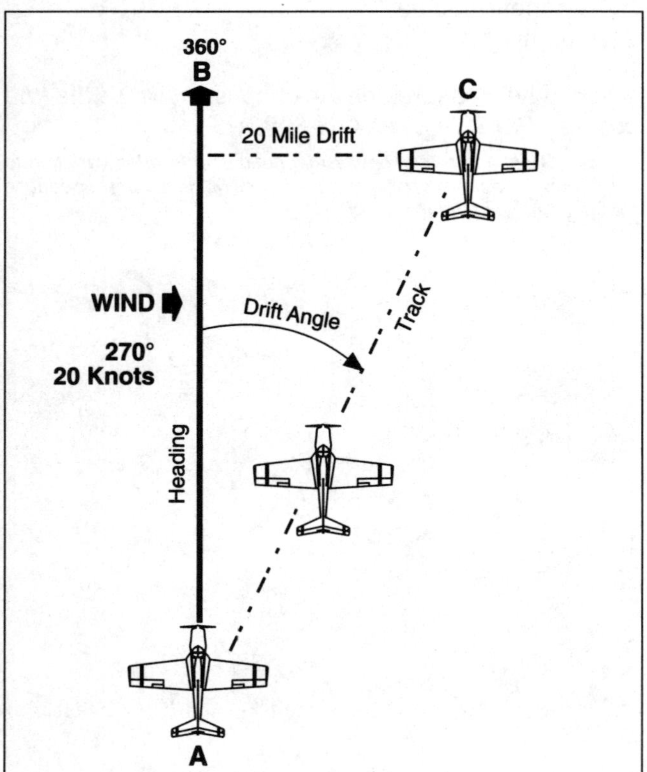

Figure 7-4. Drift

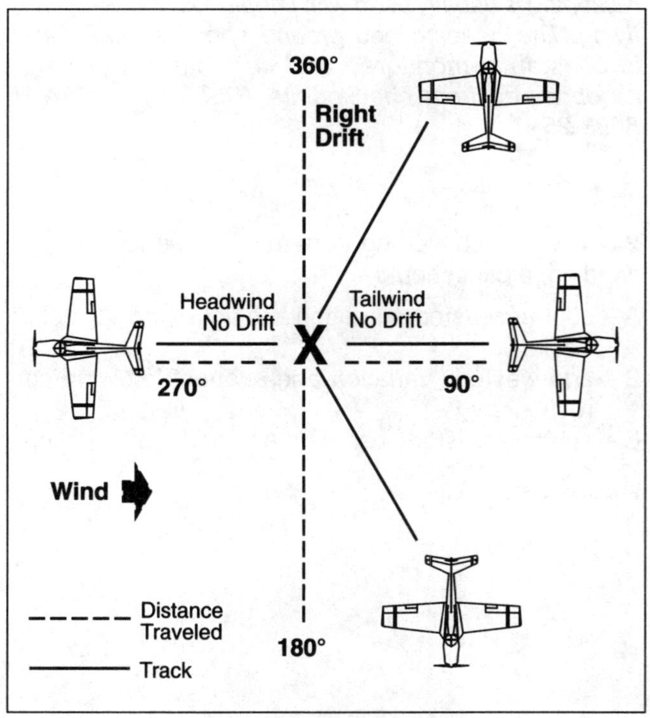

Figure 7-5. Different effects of wind on track and ground speed

A given wind will cause a different drift on each aircraft heading and it will also affect the distance traveled over the ground in a given time. With a given wind, the ground speed varies with each different aircraft heading. Figure 7-5 illustrates how a wind of 270° at 20 knots would affect the ground speed and track of an aircraft on various headings. In this particular illustration, on a heading of 360°, drift would be to the right; on a heading of 180°, drift would be to the left. With a 90° crosswind and no heading correction applied, there would be little effect on ground speed. On a heading of 090°, no left or right drift would be experienced; instead, the ground speed would be increased by the tailwind. On a 270° heading, the ground speed would be reduced by the headwind.

By determining the amount of drift, a pilot can counteract the effect of wind and make the track of the aircraft coincide with the desired course. For example, if the wind is from the left, the correction would be made by pointing the aircraft to the left a certain number of degrees, thereby compensating for wind drift. This is the Wind Correction Angle (WCA), and it is expressed as degrees left or right of the course. *See* Figure 7-6.

Any course, true, magnetic, or compass, becomes a heading when it is corrected for wind. *See* Figure 7-7 on the next page.

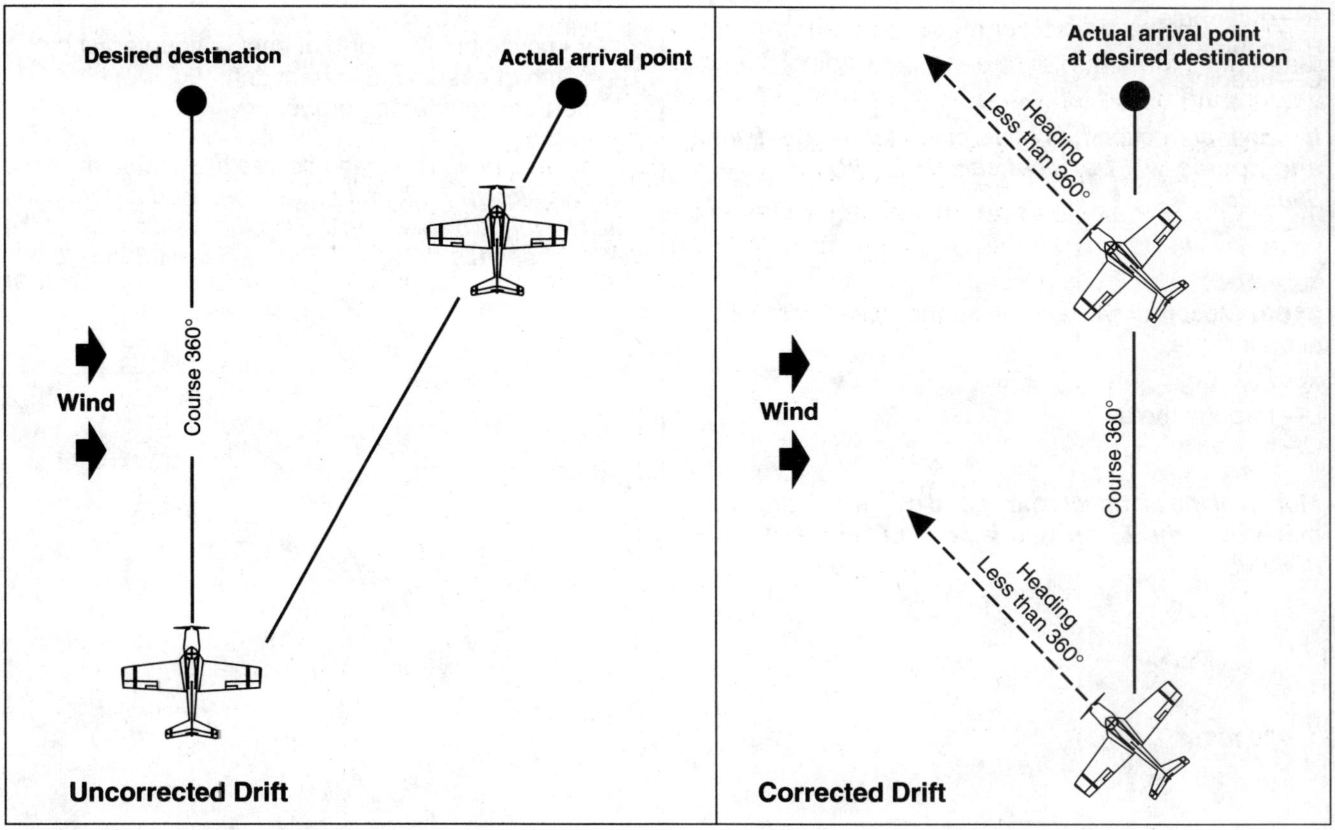

Figure 7-6. Compensating for wind drift

Course				Heading
True	±	WCA	=	True
±				±
Variation				Variation
=				=
Magnetic	±	WCA	=	Magnetic
±				±
Deviation				Deviation
=				=
Compass	±	WCA	=	Compass

Figure 7-7

ALL

2177. If proper correction has been made for the wind, ground track will be identical with

A—drift angle.
B—course.
C—heading.

If proper correction has been made for the wind, track and course will be identical. (PLT198) — FAA-H-8083-25

ALL

2103. Motion of the air affects the speed with which aircraft move

A—over the Earth's surface.
B—through the air.
C—in a turn.

Motion of the air affects the speed with which airplanes move over the Earth's surface. (PLT198) — FAA-H-8083-25

ALL

2104. If a flight is to be made on a course to the east, with a wind blowing from northeast, the aircraft must be headed

A—somewhat to the north of east to counteract drift.
B—south of east to counteract drift.
C—north to counteract torque.

If the mass of air is moving across the course from the left, the airplane will drift to the right, and a correction must be made by heading the airplane sufficiently to the left to offset this drift. (PLT198) — FAA-H-8083-25

Answers

2177 [B] 2103 [A] 2104 [A]

Flight Planning Calculations

Before a cross-country flight, a pilot should make common calculations for time, speed, and distance, and the amount of fuel required.

Converting Minutes to Equivalent Hours

It frequently is necessary to convert minutes into equivalent hours when solving speed, time, and distance problems. To convert minutes to hours, divide by 60 (60 minutes = 1 hour). Thus, 30 minutes 30/60 = 0.5 hour. To convert hours to minutes, multiply by 60. Thus, 0.75 hour equals 0.75 × 60 = 45 minutes.

Time T = D ÷ GS

To find the time (T) in flight, divide the distance (D) by the ground speed (GS). The time to fly 105 statute miles at a ground speed of 70 MPH is 105 divided by 70, or 1.5 hours. (The 0.5 hour multiplied by 60 minutes equals 30 minutes.) Answer: 1:30.

Distance D = GS × T

To find the distance flown in a given time, multiply ground speed by time. The distance flown in 1 hour 45 minutes at a ground speed of 60 MPH is 60 × 1.75, or 105 statute miles.

Ground Speed GS = D ÷ T

To find the ground speed, divide the distance flown by the time required. If an airplane flies 180 statute miles in 3 hours, the ground speed is 180 divided by 3 = 60 MPH.

Fuel Consumption FC = FR × T

To find the Fuel consumed for a time period, multiply the rate of fuel consumed per hour by the time. If the fuel consumption is 5 gallons per hour and the time flown is 2.5 hours, the fuel consumed in that time period is 5 × 2.5 hours or 12.5 gallons.

True Airspeed TAS = CAS × DF

To find the True airspeed, multiply the Calibrated airspeed (which is the Indicated airspeed corrected for instrument errors) by the density factor to get the true airspeed. This is a more complex calculation and flight computers are used to simplify it.

Density Altitude DAlt = PAlt Corrected for Temperature

To find the Density altitude, you start with a known Pressure altitude and correct it for actual temperature. This is a more complex calculation and flight computers are used to simplify it.

Flight Computers

Flight computers are used for precise calculations, mainly during flight planning. GPS is commonly used for higher precision and ease of use for inflight calculations. Most slower-speed flight planning calculations for the Sport Pilot Knowledge Exam can be calculated through the above formulas and common sense.

CX-2 Pathfinder Electronic Flight Computer

The CX-2 goes directly to the calculator mode when the computer is turned on. The main menu items can be accessed by pressing the buttons along the top of the computer. Most of the calculations presented here will be in the FLIGHT menu, which is the top left-hand key on the calculator. *See* Figure 7-8.

As shown from the keypad and the function chart, the computer performs numerous conversions and functions. We will perform some calculations in detail, then show the proper functions where the computer prompts for the inputs. *See* Figure 7-9.

The mechanical E6-B flight computer operating procedures are presented here for precise flight planning calculations. ASA's electronic flight computer, the CX-2 Pathfinder, can be used in place of the E6-B for these calculations. The CX-2 has menus set up to take your input in the same flow that your flight is planned and executed, so it is the computer doing the figuring for you, rather than a mechanical calculator or you doing them in your head. It has functions for solving all the flight planning problems, as well as standard mathematical equations. *See* Figure 7-10.

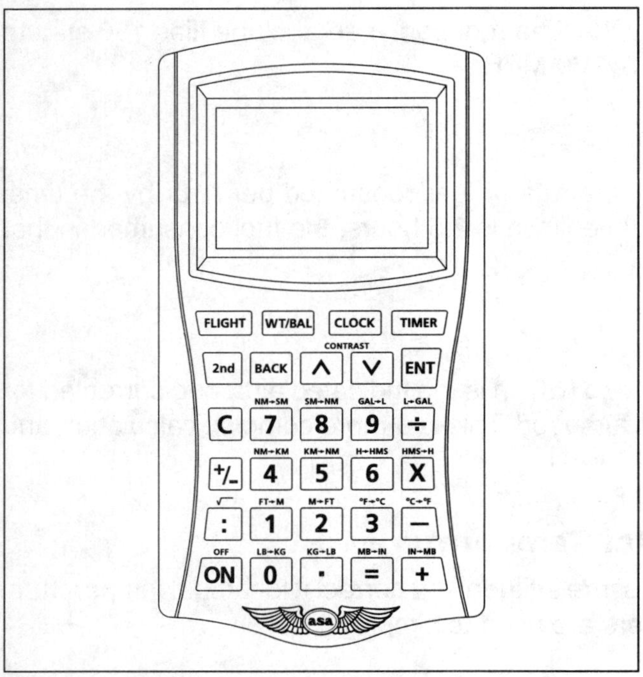

Figure 7-8. The CX-2 flight computer

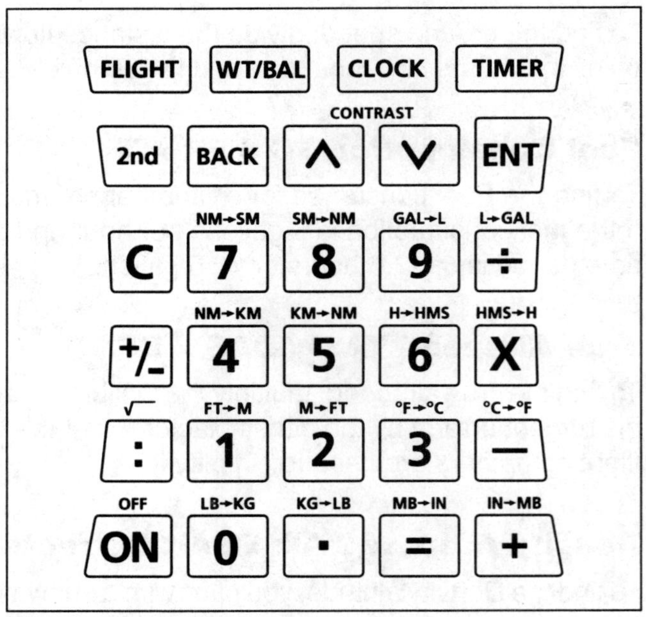

Figure 7-9. CX-2 keypad

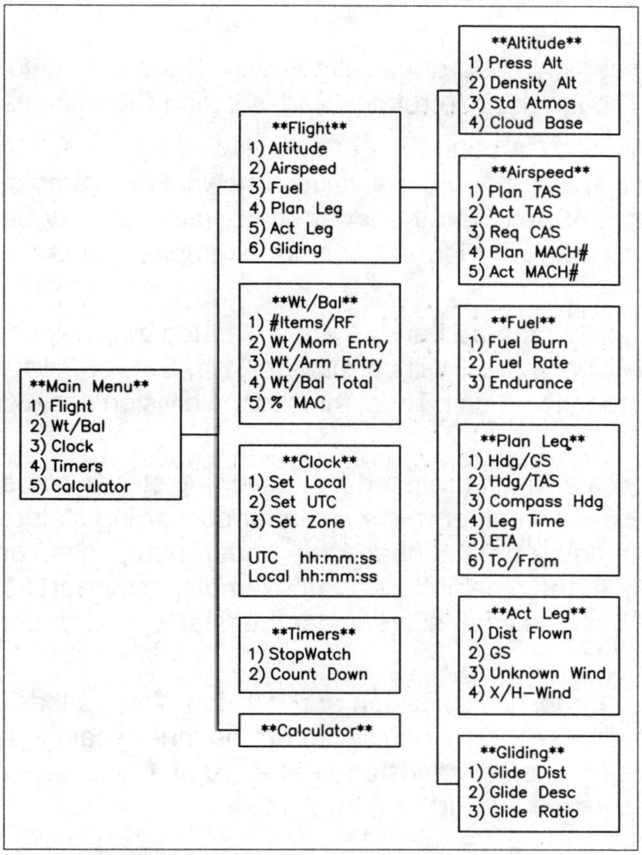

Figure 7-10. All functions are within 3 clicks. The calculator is organized for easy navigation and follows the logic associated with flight planning of a typical flight.

E6-B Circular Slide Rule

Opposite the windface of the flight computer is a circular slide rule or calculator face. The outer scale, called the miles scale, is stationary (Figure 7-11a). The inner circle rotates and is called the minutes scale (Figure 7-11b).

The numbers on the computer scale represent multiples of 10, of the values shown. For example, the number 24 on either scale may be 0.24, 2.4, 24, 240, etc. On the inner scale, minutes may be converted to hours by reference to the adjacent hours scale. In Figure 7-11c, for example, 4 hours is found adjacent to 24, meaning 240 minutes.

Relative values must be kept in mind. For example, the numbers 21 and 22 are separated by divisions, each space representing 2 units. Thus, the second division past 21 will be read as 21.4, 214, or 2,140. Between 80 and 90 are 10 divisions, each space representing 1 unit. Thus, the second division past 80 will be read as 8.2, 82 or 820.

There are three index marks on the miles (outer) scale which are used for converting statute miles (SM), nautical miles (NM) and kilometers (KM). The E6-B computer has a scale for converting statute miles, nautical miles, and kilometers. Place the known figure on the inner scale under "naut," "stat," or "KM," as appropriate, and read the equivalent value under the other indexes. For example, to convert 85 statute miles to kilometers and nautical miles, place 85 on the inner scale under the "stat" index. Under the "KM" index, read 137; under the "naut" index read 74.

Each scale has a 10-index used as a reference mark for multiplication and division. The 10-index on the inner scale can also be used as a rate index representing 1 hour. Also on the inner scale is a 60-index, representing 60 minutes and usually used for computation instead of the 10 or 1 hour index (Figure 7-11d) and a 36 or "seconds" index (3,600 seconds = 1 hour) (Figure 7-11e).

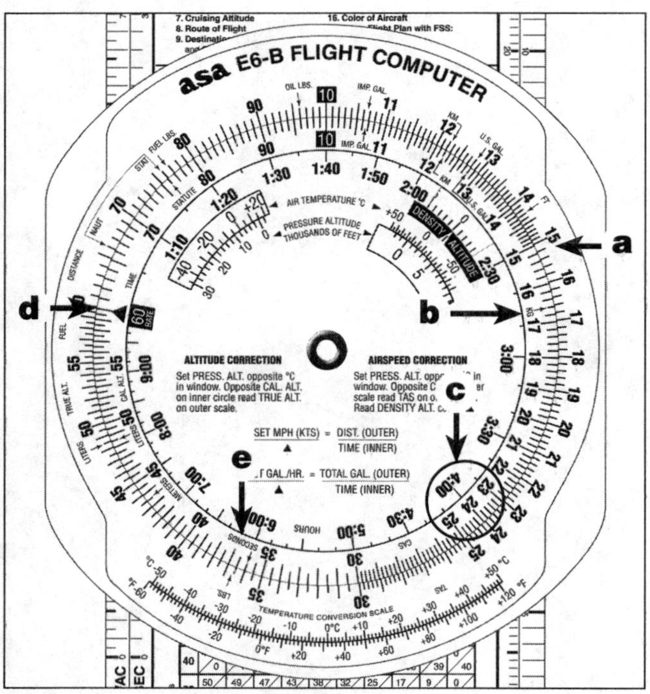

Figure 7-11. Slide rule (calculator) face

Flight Planning Problems

Calculating Time, Rate, and Distance

The flight computer will commonly be used to solve problems of time, rate, and distance. When two factors are known, the third can be found using the proportion:

$$\text{RATE (speed)} = \frac{\text{DISTANCE}}{\text{TIME}}$$

Problem: How far does an aircraft travel in 2 hours and 15 minutes at a ground speed of 138 knots?

Solution: Use the formula:

Distance = Ground Speed × Time, or use the E6-B flight computer in the following manner:

1. Place the 60 (speed) index (inner scale) under 138 (outer scale).
2. Over the time of 135 minutes or 2 hours +15 minutes (inner scale), read 310 NM (on the outer scale). *See* Figure 7-12.

Or use the CX-2 computer:

1. Press **FLIGHT** key to get to **FLIGHT** menu, press **5** to select **5. Act Leg** or arrow down to **5. Act Leg** and press **ENT**.
2. Press **1** to select **1. Dist Flown** or press **ENT** to select **1. Dist Flown**.
3. Input time **2:15**, press **ENT**.
4. Input speed **138**, press **ENT**.

The answer appears: 310 NM

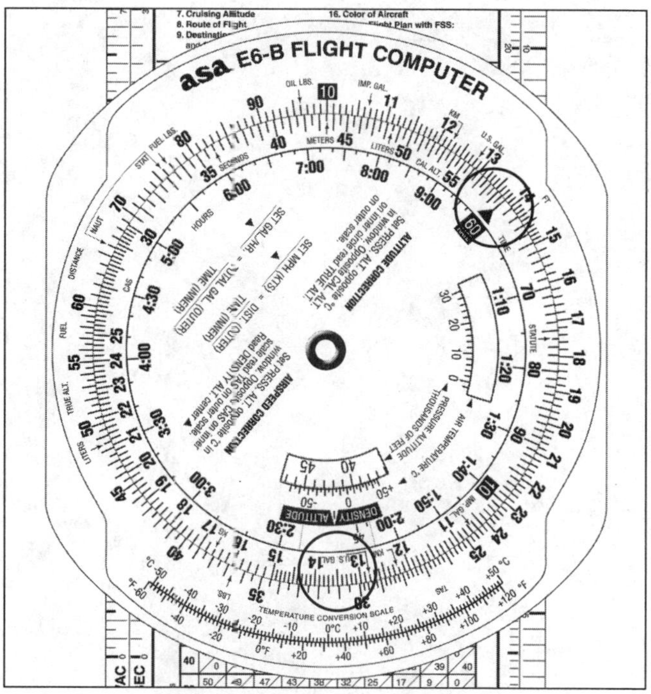

Figure 7-12. Computing distance on the E6-B

Problem: How much time is required to fly 320 nautical miles at a ground speed of 174 knots?

Solution: Use the formula:

Time = Distance ÷ Ground Speed, or use the E6-B flight computer in the following manner:

1. Set the 60 (speed) index under 174 (outer scale).
2. Under 320 (outer scale), read 110 minutes or 1 hour 50 minutes (inner scale). *See* Figure 7-13.

Or use the CX-2 calculator:

1. Press **FLIGHT** key, press **4** for Plan Leg or arrow down to **4. Plan Leg**, and press **ENT**.
2. Press **4** to select **4. Leg Time**, or arrow down to **4. Leg Time** and press **ENT**.
3. Dist input **320** NM, press **ENT**.
4. GS input **138** kts ground speed, and press **ENT**.

The answer appears: 1 hour, 50 minutes

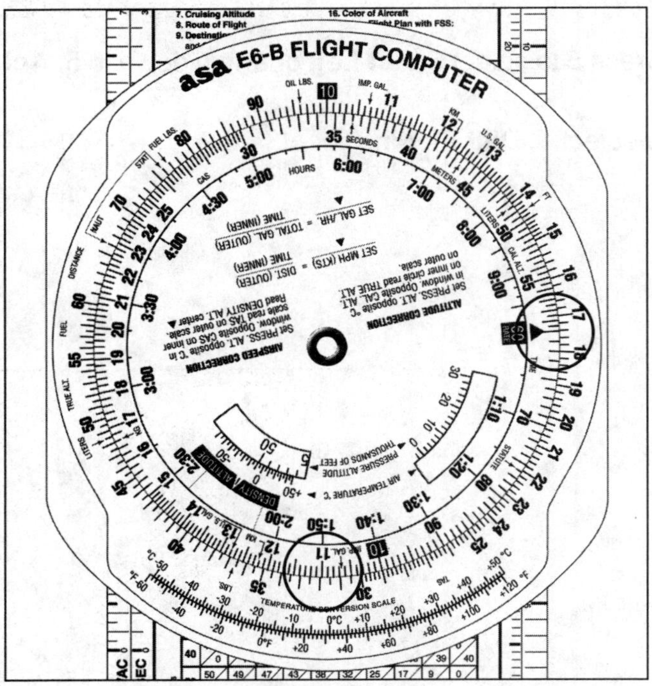

Figure 7-13. Determining time required on the E6-B

Problem: What is the ground speed if it takes 40 minutes to fly 96 nautical miles?

Solution: Use the formula:

Ground Speed = Distance ÷ Time, or use the E6-B flight computer in the following manner:

1. Place 40 minutes (inner scale) under 96 (outer scale).
2. Over the 60 (speed) index, read 144 knots (outer scale). *See* Figure 7-14.

Or use the CX-2 calculator:

1. Press **FLIGHT**, press **5** to select **5. Act Leg** or arrow down to **5. Act Leg**, and press **ENT**.
2. Press **2** for ground speed, or arrow down to **2. GS** and press **ENT**.
3. Duration input **0:40** minutes time, and press **ENT**.
4. Dist input **96** NM, and press **ENT**.

The answer appears: 144 knots

Problem: If 50 minutes are required to fly 120 nautical miles, how many minutes are required to fly 86 nautical miles at the same rate?

Solution:

1. Set 50 (inner scale) under 120 (outer scale).
2. Under 86 (outer scale), read 36 minutes required (inner scale). *See* Figure 7-15.

Or use the CX-2 calculator (or calculate from formulas):

Calculate speed, then multiply speed by time to get minutes.

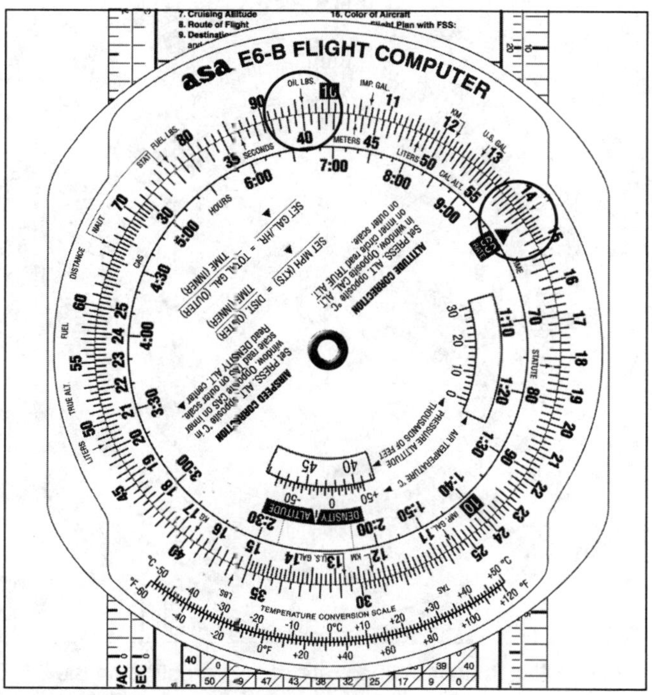

Figure 7-14. Determining ground speed on the E6-B

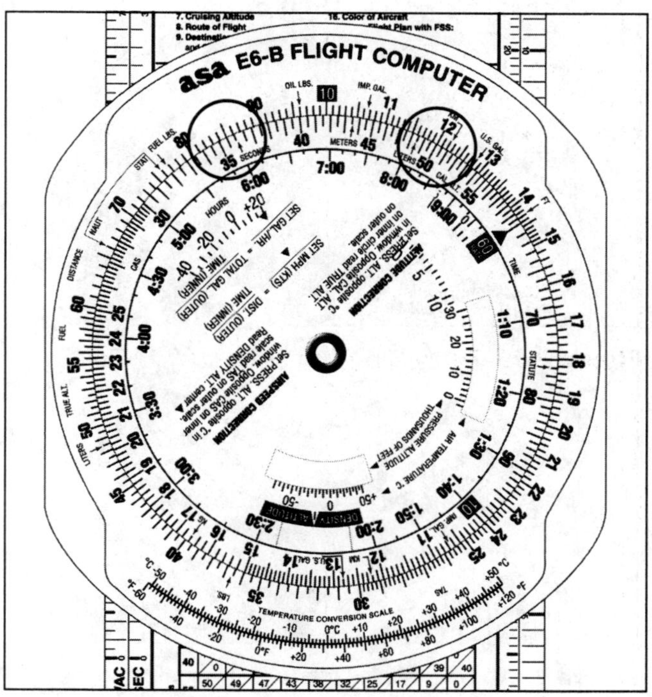

Figure 7-15. Determining time-rate-distance on the E6-B

Calculating Fuel Consumption

Fuel consumption problems are solved in the same manner as Time-Rate-Distance problems.

Problem: If 18 gallons of fuel are consumed in 1 hour, how much fuel will be used in 2 hours 20 minutes?

Solution (E6-B):

1. Set the 60-index (inner scale) under 18 (outer scale).
2. Over 2 hour 20 minute (inner scale), read 42 gallons (outer scale). *See* Figure 7-16.

Or use the CX-2 calculator:

FLIGHT menu, **3. Fuel**, **1. Fuel Burn**, input 2:20, input rate of 18 gallons per hour given, press **ENT** to read 42 gallons burned for the answer.

Problem: What is the rate of fuel consumption if 30 gallons of fuel are consumed in 111 minutes?

Solution (E6-B):

1. Set 111 (inner scale) under 30 (outer scale).
2. Over the 60-index (inner scale), read 16.2 gallons per hour (outer scale). *See* Figure 7-17.

Or use the CX-2 calculator:

FLIGHT menu, **3. Fuel**, **2. Fuel Rate**, input 111 minutes divide by 60 **ENT** to get proper hours (1:51:00), **ENT**, input 30 gallons, press **ENT** to read answer of 16.2 gallons per hour burn rate.

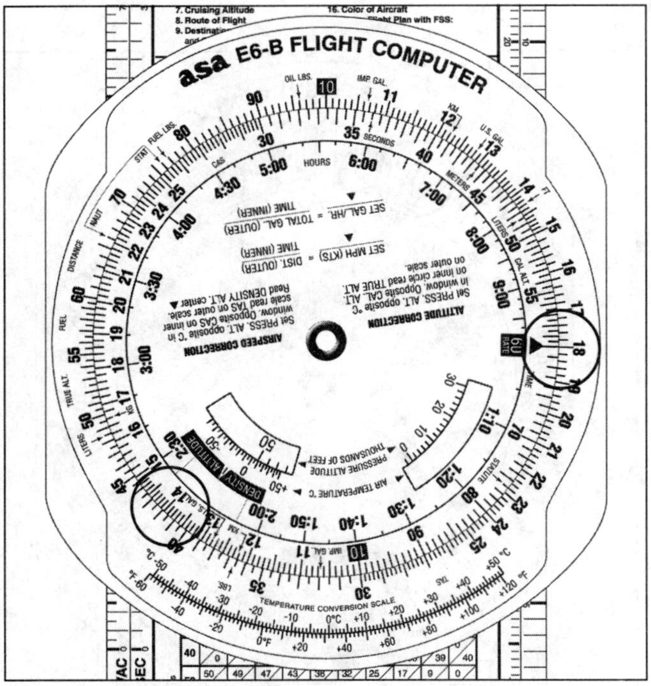

Figure 7-16. Determining total fuel consumed on the E6-B

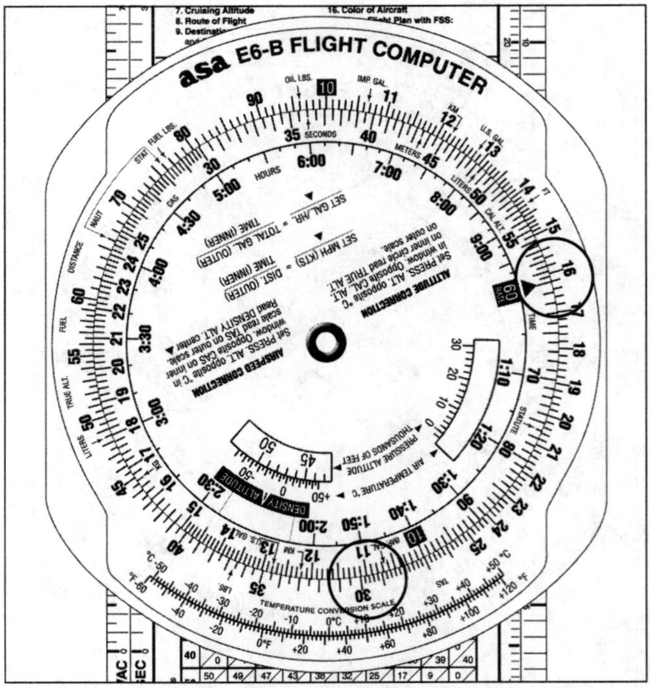

Figure 7-17. Determining rate of consumption on the E6-B

Problem: Forty gallons of fuel have been consumed in 135 minutes (2 hours and 15 minutes) flying time. How much longer can the aircraft continue to fly if 25 gallons of available fuel remain and the rate of consumption remains the same?

Solution (E6-B):

1. Set 135 (inner scale) under 40 (outer scale).
2. Under 25 (outer scale), read 84.5 minutes of fuel remaining. *See* Figure 7-18.

Or use the CX-2 calculator:

FLIGHT menu, **3. Fuel**, **2. Fuel Rate**, input 2:15 **ENT**, input 40 **ENT** to get fuel rate of 17.9 gallons per hour, press **BACK**, **3. Endurance**, input 25 gallons **ENT**, **ENT** again (since it remembers the rate) to get 1 hour 24 minutes for the answer.

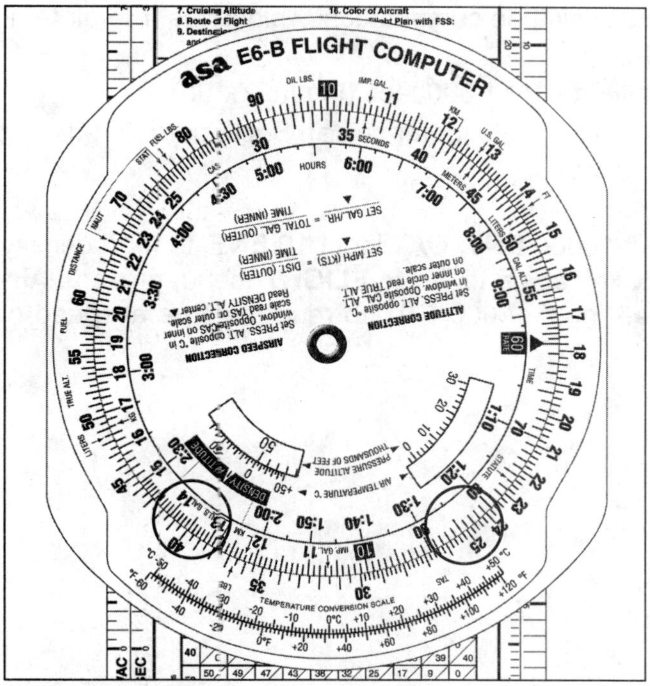

Figure 7-18. Determining flight time remaining on the E6-B

Calculating True Airspeed and Density Altitude

The calculator (slide rule) side of the computer may also be used to determine true airspeed by correcting calibrated airspeed for temperature and pressure altitude. Density altitude is computed simultaneously.

Problem: Using a flight computer and the following conditions, find the true airspeed (TAS) and the density altitude (DA).

Conditions:

Outside Air Temperature.................... +10°C
Pressure Altitude (PA) .... 10,000 feet
Calibrated Airspeed (CAS) ....140 knots

Solution using the E6-B:

1. Using the window marked "for airspeed and density altitude computations," place the true air temperature over the pressure altitude (Figure 7-19a).
2. In the density altitude window read the density altitude of 11,800 feet. (Figure 7-19b).
3. Over 140 knots CAS (inner scale) read the TAS of 167 knots (outer scale). *See* Figure 7-19c.

Or use the CX-2 calculator:

FLIGHT menu, **1. Altitude, 2. Density Alt, PAlt** input 10000 **ENT**, **OAT** input 10 **ENT**, to read density altitude of 11699. Press **BACK** key once, and **BACK** key again to get to **FLIGHT** menu, press **2. Airspeed, CAS** input 140 **ENT**, **OAT** input 10 **ENT**, **Palt** input 10000 **ENT** to read the True airspeed of 166.9 knots.

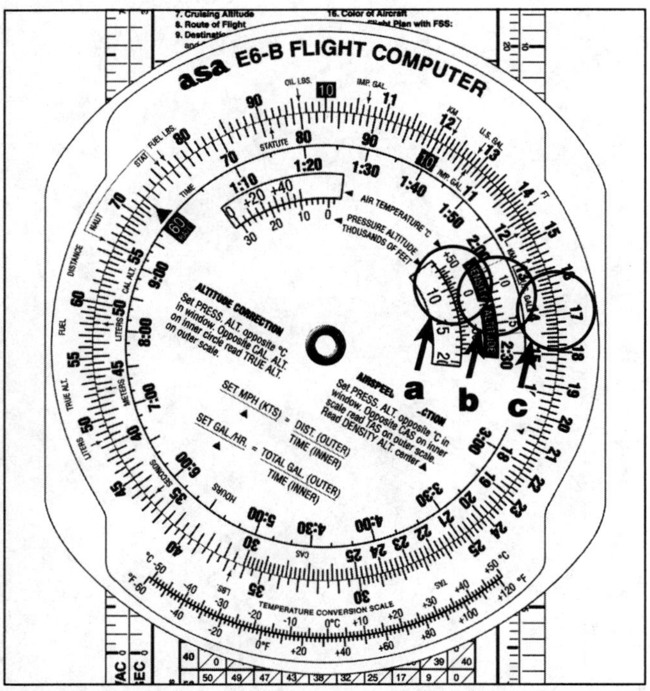

Figure 7-19. Determining true airspeed and density altitude on the E6-B

ALL

2099. If an aircraft is consuming 3 gallons of fuel per hour at a cruising altitude of 500 feet and the groundspeed is 45 mph, how much fuel is required to travel 75 SM?

A—6 gallons.
B—5 gallons.
C—3 gallons.

To determine the fuel required for a given flight, the time required for the flight must be known. Time in flight multiplied by rate of consumption gives the quantity of fuel required. For example, a flight of 75 SM at a ground speed of 45 mph requires 1.67 hours (distance divided by ground speed = time). If the plane consumes 3 gallons an hour, the total consumption will be 1.67 x 3, or 5 gallons. (PLT012) — FAA-H-8083-25

ALL

2105. To find the distance flown in a given time, multiply time by

A—groundspeed.
B—indicated airspeed.
C—equivalent airspeed.

Distance D = GS x T: To find the distance flown in a given time, multiply ground speed by time. (PLT012) — FAA-H-8083-25

GLI

2357. (Refer to Figure 59.) If a glider is launched over Barnes County Airport (area 6) with sufficient altitude to glide to Jamestown Airport (area 4), how long will it take for the flight at an average of 40 MPH groundspeed?

A—20 minutes.
B—27 minutes.
C—48 minutes.

Use the following steps:

1. *Draw a direct course line between Barnes County Airport and Jamestown Airport and determine the distance (31 SM).*
2. *Calculate the time required to cover 31 SM at 40 MPH ground speed using the formula:*

   Time = Distance ÷ Speed = 31 ÷ 40 = 0.75 hours = 46.5 minutes

(PLT012) — FAA-H-8083-25

GLI

2358. (Refer to Figure 60, area 1.) A glider is launched over Caddo Mills Airport with sufficient altitude to glide to Airpark East Airport, south of Caddo Mills. How long will it take for the flight at an average of 35 MPH groundspeed?

A—27 minutes.
B—25 minutes.
C—31 minutes.

Use the following steps:

1. *Draw a direct course line between Caddo Mills Airport and Airpark East Airport and determine the distance (18 SM).*
2. *Calculate the time required to cover 18 SM at 35 MPH ground speed using the formula:*

   Time = Distance ÷ Speed = 15 ÷ 35 = .514 hours = 31 minutes

(PLT012) — FAA-H-8083-25

Answers

2099 [B] 2105 [A] 2357 [C] 2358 [C]

Finding Wind Correction Angle (WCA) and Ground Speed

The wind correction angle required to change a course to a heading can be found by using the wind face of the E6-B circular slide rule or the CX-2 electronic flight computer. Ground speed can also be determined using this procedure.

The wind face of the E6-B consists of a transparent, rotatable plotting disk mounted in a frame as shown in Figure 7-20. A compass rose is located around the plotting disk (Figure 7-20a). A correction scale on the top of the frame is graduated in degrees left and right of the true index (Figure 7-20b), and it is used for calculating drift correction. A small reference circle, called a grommet, is located at the center of the plotting disk (Figure 7-20c).

A sliding grid (Figure 7-20d) inserted behind the plotting disk is used for wind computations. The slide has converging lines (track or drift lines) which indicate degrees left or right of center (Figure 7-20e). Concentric arcs (speed circles) on the slide are used for calculations of speed (Figure 7-20f). The wind correction angle and ground speed may be found when true course, true airspeed, wind speed and wind direction (relative to true north) are known.

Problem: Using a flight computer and the following conditions, find the wind correction angle (WCA), the true heading (TH), and the ground speed (GS).

Conditions:

True course (TC)....................................090°

True airspeed (TAS)......................120 knots

True wind direction...............................160°

Wind speed....................................30 knots

Solution:

1. Using the wind face side of the E6-B, set the true wind direction (160°) under the true index (Figure 7-21a).
2. With the sliding grid set at some arbitrary value, say 160 at the grommet, measure above the grommet 30 units (wind speed) and put a wind dot (pencil mark) at this point (Figure 7-21b).
3. Rotate the plotting disk and set the true course (090°) under the true index (Figure 7-22a).
4. Adjust the sliding grid so that the TAS arc (120 knots) is at the wind dot (Figure 7-22b). Note the wind dot is 14° right of centerline (Figure 7-22c), so the WCA = 14°R.
5. Under the grommet, read the ground speed (GS) of 106 knots (Figure 7-22d).
6. Determine the true heading by applying the formula:

   TC + WCA = TH

   In this case the wind correction is to the right, so it will be added to the true course:

090°	(TC)
+ 14°	(WCA)
104°	(TH)

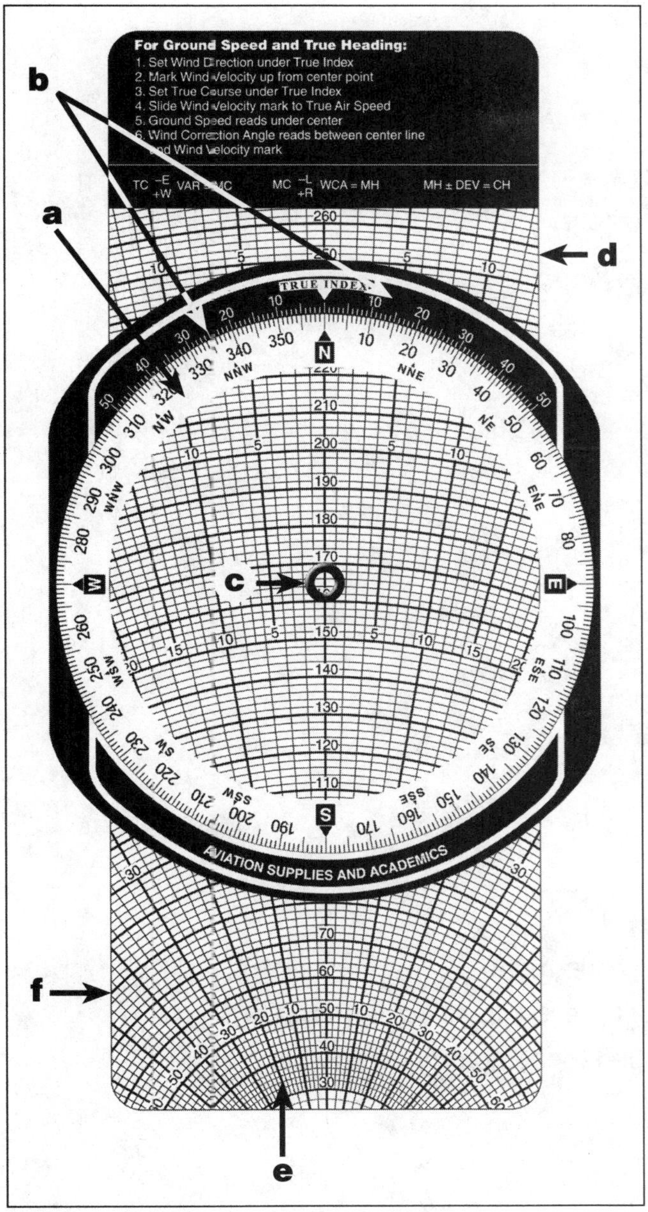

Figure 7-20. Wind face of E6-B computer

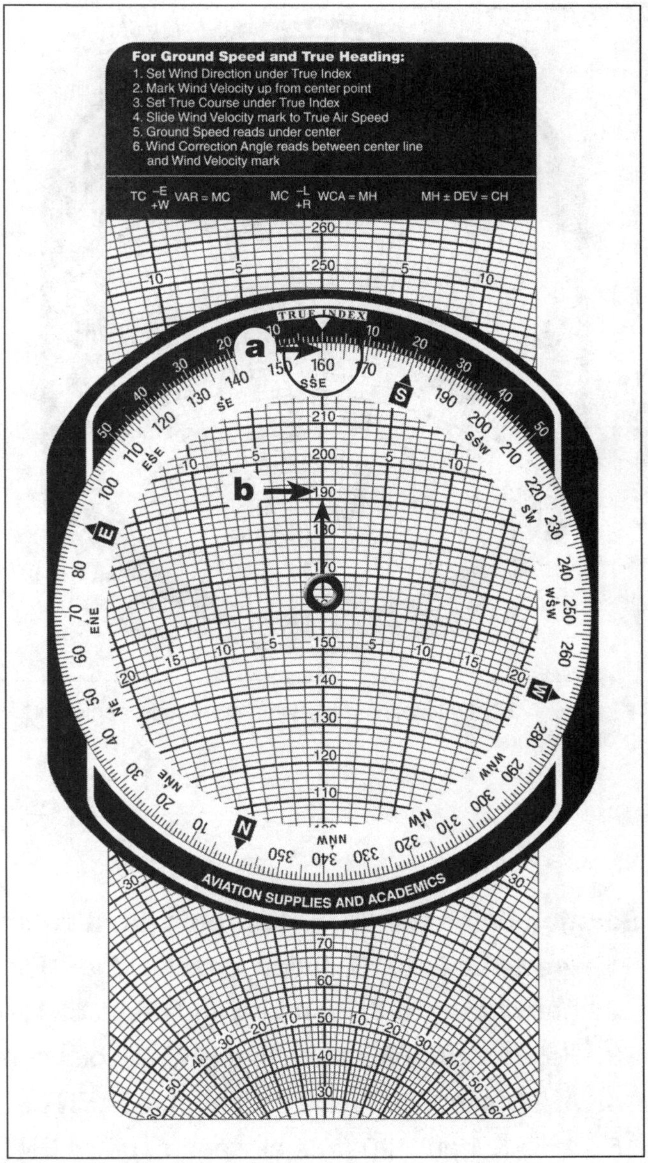

Figure 7-21. Plotting wind speed and wind direction

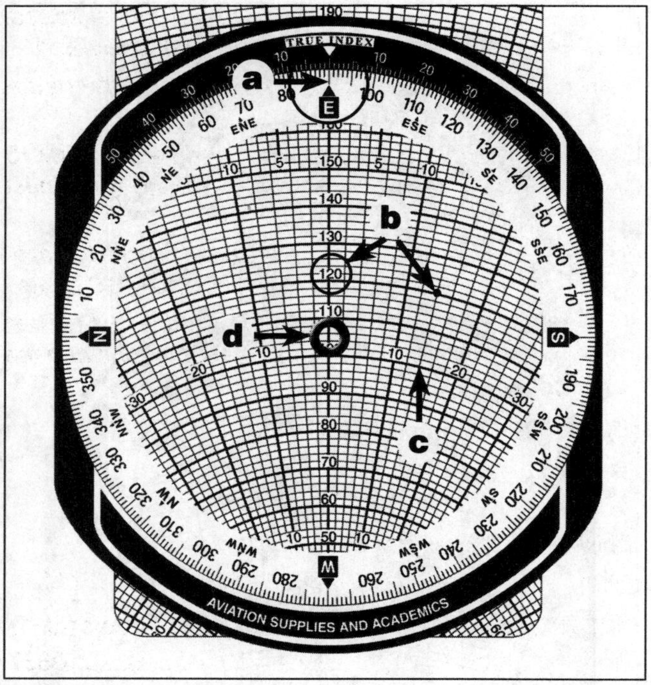

Figure 7-22. Determining the wind correction angle and ground speed

Solution using the CX-2 electronic flight computer:

1. Press the **FLIGHT** button, or select the **Flight** main menu.
2. Arrow down to **4. Plan Leg** and press **ENT**, or just press 4 to select **4. Plan Leg**.
3. Select **1.Hdg/GS** since that is what you want to find (heading/ground speed).
4. At **TCrs** input 90 (true course), press **ENT**.
5. At **TAS** input 120 (true air speed), press **ENT**.
6. At **Wdir** input 160 (wind direction), press **ENT**.
7. At **Wspd** input 30 (wind speed), press **ENT**.

The answer is displayed:

THdg 103.6 (true heading)

GS 106.4 (ground speed)

ALL

2101. Given:

True course..050
True Heading...040
True airspeed...75 kts
Groundspeed..65 kts

Determine the wind direction and speed.

A—105°; and 16 knots
B—355°; and 16 knots
C—355°; and 10 knots

Using an E6-B computer:

1. *Turn compass azimuth so 050° is at the true index.*
2. *Slide board so grommet is on the ground speed arc, 65 knots.*
3. *Determine the wind correction angle (WCA) by comparing course to heading.*

    050° – 040° = 10° left
4. *Plot WCA at TAS arc with a pencil dot, 75 knots.*
5. *Rotate compass azimuth so pencil dot is on the centerline.*
6. *Read wind direction under true index and wind speed as number of units between grommet and pencil dot, 354° and 16 knots.*

Alternate method, using CX-2 computer:

1. *Go to FLIGHT menu, select 5. Act Leg.*
2. *Select 3. Unknown Wind, and enter 40 for THdg input, press ENT.*
3. *GS input 65 knots, press ENT.*
4. *TCrs input 50 for true course, press ENT.*
5. *TAS input 75 for true airspeed, press ENT.*
6. *Read unknown wind direction of 354° at 15.8 kts.*

Alternate method: using logic to find answer.

1. *Picture your true course flying northeast.*
2. *Picture your heading 10 degrees to the north.*
3. *This indicates a crosswind from your left.*

Continued

4. *Note your ground speed is lower than your airspeed. This indicates a headwind rather than a tailwind.*
5. *From items 3 and 4 above, the wind can be determined from the northwest.*
6. *Your true airspeed and ground speed are exactly 10 knots apart; since you are crabbed, the wind must be stronger than 10 knots overall.*
7. *You pick answer B because the wind is from the estimated direction, and the speed is above 10 knots.*
8. *Looking at the answer and determining direction and speed, this confirms your assumptions and the correct answer.*

(PLT012) — FAA-H-8083-25

LTA

2222. (Refer to Figure 61, point 1.)

Given:

Departure point..................Georgetown Airport (Q61)
Departure time..0637
Winds aloft forecast (FD) at your altitude 1008
Airspeed ....8 kts

At 0755, the aircraft should be

A—over Auburn Airport (AUN).
B—over the town of Auburn.
C—slightly west of the town of Garden Valley.

1. *Calculate time en route:*

    0755 – 0637 = 1:18 = 1.3 hours
2. *Compute distance traveled:*

    Distance = Speed x Time

    8 x 1.3 = 10.4 NM
3. *Plot course of 100° + 180° = 280° out to a distance of 10.4 NM from Georgetown Airport (Q61), point 1. The aircraft should be just south of there.*

(PLT012) — FAA-H-8083-25

Answers

2101 [B] 2222 [A]

VFR Flight Plans

Filing a flight plan is not required by regulations; however, it is a good operating practice, since the information contained in the flight plan can be used in search and rescue in the event of an emergency. Flight plans can be filed in the air by radio, but it is best to file a flight plan either in person at the FSS or by phone just before departing. After takeoff, contact the FSS by radio and give them the takeoff time so the flight plan can be activated. When a VFR flight plan is filed, it will be held by the FSS until 1 hour after the proposed departure time and then canceled unless: the actual departure time is received; or a revised proposed departure time is received; or at the time of filing, the FSS is informed that the proposed departure time will be met, but actual time cannot be given because of inadequate communication. The FSS specialist who accepts the flight plan will not inform the pilot of this procedure, however. The pilot must close a VFR flight plan at the completion of a flight. This can be done by contacting the nearest flight service station (FSS) or other FAA facility upon landing.

ALL

2175. Why should you file a VFR flight plan with FSS?

A—You are required to do so when flying beyond 25 NM from an airport.

B—To provide traffic reports with Flight Service.

C—To provide search and rescue in the event of an emergency.

Filing a flight plan is not required by regulations; however, it is a good operating practice, since the information contained in the flight plan can be used in search and rescue in the event of an emergency. (PLT225) — FAA-H-8083-25

Answers

2175 [C]

Cross-Reference A
Question Number and Page Number

The following list of the numbered questions included in this ASA Test Prep is given in sequential order; however, as a result of our ongoing review of FAA test question databases, some question numbers may have been removed due to changes in the database. **All currently existing questions are accounted for in this list.** For more information about the questions included in ASA Test Preps, please read Pages v and vi in the front matter for this book.

Question Number	Page Number
2001	6–11
2002	5–6
2003	6–12
2004	6–12
2005	6–12
2006	6–12
2007	6–12
2008	4–14
2009	4–14
2010	4–15
2011	4–13
2012	6–14
2013	6–12
2014	6–13
2015	6–13
2016	6–13
2017	6–13
2018	5–12
2019	5–10
2020	6–6
2021	5–8
2022	5–8
2023	5–8
2024	5–8
2025	5–8
2026	5–8
2027	5–9
2028	5–9
2029	5–9
2030	5–9
2031	5–9
2032	5–9
2033	6–4
2033-1	1–17
2034	6–13
2035	5–5
2036	5–5
2037	5–5
2038	1–23
2039	6–19
2040	1–13
2041	5–17
2042	5–17
2043	5–18
2044	5–18
2045	5–18
2046	5–19
2047	5–19
2048	5–19
2049	5–19
2050	5–19
2051	5–20
2052	5–20
2053	1–25
2054	6–15
2055	5–20
2056	5–20
2057	5–20
2058	5–20
2059	5–21
2060	5–5
2061	1–6
2062	6–17
2063	3–13
2064	6–17
2065	6–16
2066	6–16
2067	4–13
2068	3–13
2069	3–13
2070	6–19
2071	6–19
2072	6–19
2073	3–10
2074	3–10
2075	3–10
2076	3–10
2077	3–11
2078	1–7
2079	3–11
2080	3–11
2081	3–13
2082	3–14
2083	3–11
2084	3–12
2085	3–10
2086	3–10
2087	1–9
2088	6–16
2089	3–4
2090	3–3
2091	3–4
2092	3–4
2093	3–4
2094	3–5
2095	3–5
2096	6–17
2097	3–5
2098	3–6
2099	7–23
2100	7–4
2101	7–27
2102	7–9
2103	7–12
2104	7–12
2105	7–23
2106	7–4
2107	1–23
2108	7–4
2109	1–23
2110	1–23
2111	5–21
2112	5–21
2113	5–22
2114	5–21
2115	5–22
2116	5–21
2117	1–24
2118	1–24
2119	1–24
2120	1–24
2121	1–24
2122	1–24
2123	1–6

Question Number	Page Number
2124	1–3
2125	1–25
2126	1–9
2127	1–9
2128	1–9
2129	1–7
2130	1–6
2131	1–6
2132	6–3
2133	1–10
2134	1–11
2135	1–11
2136	1–12
2137	1–13
2138	5–22
2139	1–9
2140	1–15
2141	1–7
2142	1–8
2143	1–17
2144	1–17
2145	1–18
2146	1–18
2147	6–4
2148	1–20
2149	1–20
2150	1–20
2151	1–20
2152	1–20
2153	2–21
2154	2–21
2155	2–19
2156	2–19
2157	2–20
2158	2–13
2159	2–11
2160	2–11
2162	2–15
2163	2–15
2164	2–7
2165	2–7
2166	2–9
2167	2–7
2168	4–14
2169	2–6
2170	2–5
2171	2–3
2172	2–13
2173	2–13
2174	4–19
2175	7–28
2176	3–11
2177	7–12
2178	7–4
2179	5–22
2180	1–23
2181	2–3
2182	4–19
2183	4–20
2184	6–17
2185	1–6
2186	1–13
2187	1–13
2188	1–8
2189	1–20
2190	1–15
2191	2–19
2192	5–17
2193	5–18
2194	5–6
2195	3–10
2196	3–11
2197	3–13
2198	5–21
2199	4–15
2200	5–18
2201	7–9
2202	3–4
2203	1–13
2204	5–22
2205	6–11
2206	5–7
2207	2–6
2208	4–14
2209	2–19
2210	2–21
2211	1–21
2212	1–21
2213	1–21
2214	1–21
2215	4–5
2216	4–5
2217	4–13
2218	5–9
2219	5–9
2220	6–4
2221	7–9
2222	7–27
2223	4–6
2224	4–6
2225	4–5
2226	6–15
2227	4–13
2228	6–7
2229	4–5
2230	4–6
2231	6–8
2232	1–18
2233	6–8
2234	6–8
2235	6–19
2236	4–6
2237	4–5
2238	4–22
2239	4–5
2240	3–6
2241	5–19
2242	4–22
2243	4–23
2244	4–23
2245	4–23
2246	4–23
2247	4–23
2248	4–23
2249	4–24
2250	4–24
2251	4–24
2252	4–24
2253	4–24
2254	4–25
2255	4–25
2256	4–25
2257	4–25
2258	4–25
2259	4–25
2260	4–26
2261	4–26
2262	4–26
2263	4–27
2264	4–27
2265	4–27
2266	4–28
2267	4–28
2268	4–28
2269	4–28
2270	4–28
2271	4–29
2272	4–29
2273	4–29
2274	4–29
2275	4–29
2276	4–29

Cross-Reference B

Learning Statement Code and Question Number

The expression "learning statement," as used in FAA airman testing, refers to measurable statements about the knowledge a student should be able to demonstrate following a certain segment of training. In order that each learning statement may be read and understood in context as a complete sentence, precede each statement with the words: "Upon the successful completion of training the student should be able to..."—complete the phrase with the subject indicated by the learning statement code (LSC) given in your knowledge test results.

When you take the applicable airman knowledge test required for an airman pilot certificate or rating, you will receive an Airman Knowledge Test Report. The test report will list the learning statement codes for questions you have answered incorrectly. Match the codes given on your test report to the ones in the official FAA Learning Statement Codes (listed below in this cross-reference). Use Cross-Reference A in this book to find the page number for the question numbers listed below. Your instructor is required to provide instruction on each of the areas of deficiency listed on your Airman Knowledge Test Report and give you an endorsement for this instruction. The Airman Knowledge Test Report must be presented to the examiner conducting your practical test. During the oral portion of the practical test, the examiner is required to evaluate the noted areas of deficiency.

If you received a code on your Airman Test Report that is not listed in this cross-reference, email ASA at **cfi@asa2fly.com**. We will provide the definition so you can review that subject area.

The FAA appreciates testing experience feedback. You can contact the branch responsible for the FAA Knowledge Exams directly at:

Federal Aviation Administration
AFS-630, Airman Testing Standards Branch
PO Box 25082
Oklahoma City, OK 73125
Email: AFS630comments@faa.gov

PLT005 .... Calculate aircraft performance—density altitude
2356

PLT011 .... Calculate aircraft performance—takeoff
2361, 2362

PLT012 .... Calculate aircraft performance—time/speed/distance/course/fuel/wind
2099, 2101, 2105, 2221, 2222, 2320, 2346, 2357, 2358

PLT015 .... Calculate flight performance/planning—range
2230

PLT017 .... Calculate L/D ratio
2359

PLT021 .... Calculate weight and balance
2350, 2351

PLT022 .... Define Aeronautical Decision Making (ADM)
2079, 2080, 2085

PLT023 .... Define altitude—absolute/true/indicated/density/pressure
2217, 2335, 2336, 2337

PLT026 .... Define ceiling
2155

PLT039 .... Interpret airport landing indicator
2036, 2037, 2324

PLT040 .... Interpret airspace classes—charts/diagrams
2041, 2042, 2045, 2050

PLT041 .... Interpret altimeter—readings/settings
2303, 2355

PLT044 .... Interpret ATC communications/instructions/terminology
2003, 2012

PLT059 .... Interpret information on a METAR/SPECI report
2156, 2209

PLT064 .... Interpret information on a Sectional Chart
2013, 2014, 2034, 2038, 2047, 2051, 2057, 2059, 2111, 2112, 2113, 2114, 2115, 2116, 2179, 2198, 2204, 2302, 2304, 2305, 2306, 2318, 2319

PLT074 .... Interpret information on a Velocity/Load Factor Chart
2215, 2216

PLT078 .... Interpret information on an Airport Facility Directory (AFD)
2107, 2119

PLT097 .... Recall aeromedical factors—effects of carbon monoxide poisoning
2095

PLT098 .... Recall aeromedical factors—fitness for flight
2089, 2090, 2202

PLT099 .... Recall aeromedical factors—scanning procedures
2096

PLT103 .... Recall Aeronautical Decision Making (ADM)—hazardous attitudes
2074, 2075, 2076, 2077, 2176, 2195, 2196

PLT104 .... Recall Aeronautical Decision Making (ADM)—human factors
2081, 2082, 2083, 2084, 2197

PLT112 .... Recall aircraft controls—proper use/techniques
2015, 2228

PLT114 .... Recall aircraft design—construction/function
2263, 2264, 2265, 2266, 2268, 2269, 2271, 2272, 2273, 2274, 2275, 2276, 2277, 2278, 2347, 2348

PLT115 .... Recall aircraft engine—detonation cause/characteristics
2344

PLT116 .... Recall aircraft general knowledge/publications/AIM/navigational aids
2004, 2019, 2052, 2062, 2066, 2106, 2108, 2109, 2110, 2117, 2180, 2226

PLT119 .... Recall aircraft lighting—anti-collision/landing/navigation
2020

PLT120 .... Recall aircraft limitations—turbulent air penetration
2311

PLT122 .... Recall aircraft operations—checklist usage
2068

PLT123 .... Recall aircraft performance—airspeed
2326

PLT124 .... Recall aircraft performance—atmospheric effects
2008, 2009, 2010

PLT125 .... Recall aircraft performance—climb/descent
2229, 2300, 2316

PLT126 .... Recall aircraft performance—cold weather operations
2353, 2354

PLT127 .... Recall aircraft performance—density altitude
2011, 2067

PLT129 .... Recall aircraft performance—effects of runway slope
2199

PLT130 .... Recall aircraft performance—fuel
2293, 2301

PLT131 .... Recall aircraft performance—ground effect
2223

PLT132 .... Recall aircraft performance—instrument markings/airspeed/definitions/indications
2334

PLT134 .... Recall aircraft performance—takeoff
2234, 2283

PLT141 .... Recall airport operations—markings/signs/lighting
2021, 2022, 2023, 2024, 2025, 2026, 2027, 2028, 2029, 2030, 2031, 2032, 2206, 2218, 2219

PLT147 .... Recall airport operations—visual glideslope indicators
2018, 2282, 2325

PLT149 .... Recall airport taxi operations—procedures
2328, 2331, 2349

PLT152 .... Recall airship—flight characteristics/controllability
2321

PLT153 .... Recall airship—flight operations
2317

PLT159 .... Recall airship—pressure height/super heat/valving gas
2314

PLT160 .... Recall airship—stability/control/positive superheat
2322

PLT161 .... Recall airspace classes—limits/requirements/restrictions/airspeeds/equipment
2044, 2046, 2048, 2049, 2200, 2241

PLT162 .... Recall airspace requirements—operations
2043, 2056, 2192, 2193

PLT163 .... Recall airspace requirements—visibility/cloud clearance
2040, 2137, 2138, 2203

PLT165 .... Recall altimeter—effect of temperature changes
2181

PLT170 .... Recall approach/landing/taxiing techniques
2001, 2002, 2017

PLT172 .... Recall ATC—system/services
2205

PLT173 .... Recall atmospheric conditions—measurements/pressure/stability
2167

PLT177 .... Recall balloon—flight operations
2295, 2309

PLT179 .... Recall balloon—ground weigh-off/static equilibrium/load
2310

PLT181 .... Recall balloon—hot air/physics
2285

PLT183 .... Recall balloon flight operations—ascent/descent
2298

PLT184 .... Recall balloon flight operations—launch/landing
2296

PLT190 .... Recall carburetor ice—factors affecting/causing
2243, 2249, 2260, 2341

PLT192 .... Recall clouds—types/formation/resulting weather
2160, 2161, 2166

PLT194 .... Recall collision avoidance—scanning techniques
2054, 2064, 2065, 2088, 2184

PLT198 .... Recall course/heading—effects of wind
2103, 2104, 2177

PLT200 .... Recall dead reckoning—calculations/charts
2100

PLT201 .... Recall departure procedures—ODP/SID
2035

PLT205 .... Recall effects of alcohol on the body
2087

PLT206 .... Recall effects of temperature—density altitude/icing
2168

PLT207 .... Recall electrical system—components/operating principles/characteristics
2238, 2254, 2333

PLT208 .... Recall emergency conditions/procedures
2315

PLT213 .... Recall flight characteristics—longitudinal stability/instability
2224

PLT214 .... Recall flight characteristics—structural/wing design
2267

PLT219 .... Recall flight operations—maneuvers
2231, 2299

PLT221 .... Recall flight operations—takeoff/landing maneuvers
2279, 2280

PLT222 .... Recall flight operations—takeoff procedures
2360

PLT225 .... Recall flight plan—requirements
2175

PLT239 .... Recall forces acting on aircraft—buoyancy/drag/gravity/thrust
2313

PLT241 .... Recall forces acting on aircraft—drag/gravity/thrust/lift
2237

PLT242 .... Recall forces acting on aircraft—lift/drag/thrust/weight/stall/limitations
2261, 2262

PLT251 .... Recall fuel characteristics/contaminants/additives/leaks
2220, 2288, 2291

PLT253 .... Recall fuel system—components/operating principles/characteristics
2244, 2245, 2246, 2248, 2259, 2289, 2338, 2342

PLT254 .... Recall fuel tank—components/operating principles/characteristics
2287, 2290, 2292

PLT257 .... Recall glider performance—speed/distance/ballast/lift/drag
2239

PLT259 .... Recall ground resonance—conditions to occur
2329, 2330

PLT260 .... Recall gyroplane—aerodynamics/rotor systems
2327

PLT267 .... Recall hot air balloon—weigh-off procedure
2284

PLT271 .... Recall human factors (ADM)—judgment
2073, 2086

PLT280 .... Recall inflight illusions—causes/sources
2098, 2240

PLT281 .... Recall information in an Airport Facility Directory
2118

PLT290 .... Recall information on AIRMETs/SIGMETs
2157

PLT291 .... Recall information on an Aviation Area Forecast (FA)
2281

PLT304 .... Recall launch procedures
2297

PLT305 .... Recall leading edge devices—types/effect/purpose/operation
2286

PLT313 .... Recall loading—limitations
2174, 2182

PLT320 .... Recall navigation—true north/magnetic north
2102

PLT323 .... Recall NOTAMS—classes/information/distribution
2120, 2121, 2122

PLT324 .... Recall oil system—types/components/functions
2256, 2257

PLT328 .... Recall performance planning—aircraft loading
2183, 2323

PLT332 .... Recall physiological factors—hyperventilation
2091, 2092, 2093, 2094

PLT334 .... Recall physiological factors—spatial disorientation
2097

PLT335 .... Recall pilotage—calculations
2178, 2201

PLT337 .... Recall pitot-static system—components/operating principles/characteristics
2339, 2340

PLT340 .... Recall positive exchange of flight controls
2069

PLT342 .... Recall powerplant—controlling engine temperature
2251

PLT343 .... Recall powerplant—operating principles/operational characteristics/inspecting
2247, 2250, 2252, 2253, 2258

PLT346 .... Recall primary flight controls—types/purpose/functionality
2270

PLT351 .... Recall propeller system—types/components/operating principles/characteristics
2227

PLT353 .... Recall Radar Summary Chart
2191

PLT366 .... Recall regulations—accident/incident reporting and preserving wreckage
2148, 2149, 2150, 2151, 2152, 2189, 2211, 2212

PLT371 .... Recall regulations—Aircraft Category/Class
2185

PLT374 .... Recall regulations—aircraft owner/operator responsibilities
2143

PLT376 .... Recall regulations—airspace special use/TFRS
2053, 2058

PLT377 .... Recall regulations—airworthiness certificates/requirements/responsibilities
2190

PLT378 .... Recall regulations—Airworthiness Directives
2125

PLT387 .... Recall regulations—change of address
2131

PLT393 .... Recall regulations—controlled/restricted airspace—requirements
2055, 2294

PLT395 .... Recall regulations—definitions
2124, 2213

PLT399 .... Recall regulations—display/inspection of licenses and certificates
2129

PLT400 .... Recall regulations—documents to be carried on aircraft during flight
2140

PLT403 .... Recall regulations—emergency deviation from regulations
2141, 2142, 2188

PLT414 .... Recall regulations—general right-of-way rules
2134, 2135

PLT416 .... Recall regulations—immediate notification
2214

PLT430 .... Recall regulations—minimum safe/flight altitude
2136

PLT431 .... Recall regulations—operating near other aircraft
2133

PLT435 .... Recall regulations—operational procedures for an uncontrolled airport
2006, 2007, 2060, 2194

PLT436 .... Recall regulations—operations manual
2063

PLT441 .... Recall regulations—pilot briefing
2232

PLT443 .... Recall regulations—pilot qualifications/privileges/responsibilities
2078

PLT444 .... Recall regulations—pilot-in-command authority/responsibility
2033-1, 2144, 2308

PLT445 .... Recall regulations—preflight requirements
2033, 2132, 2147

PLT448 .... Recall regulations—privileges/limitations of pilot certificates
2123, 2130

PLT463 .... Recall regulations—use of narcotics/drugs/intoxicating liquor
2126, 2127, 2128, 2139

PLT464 .... Recall regulations—use of safety belts/harnesses (crew member)
2146

PLT465 .... Recall regulations—use of seats/safety belts/harnesses (passenger)
2145

PLT467 .... Recall regulations—visual flight rules and limitations
2061, 2186, 2187

PLT470 .... Recall rotor system—types/components/operating principles/characteristics
2332

PLT477 .... Recall stalls—characteristics/factors/recovery/precautions
2225, 2233

PLT478 .... Recall starter/ignition system—types/components/operating principles/characteristics
2242, 2255, 2343, 2345

PLT486 .... Recall taxiing/takeoff—techniques/procedures
2016

PLT495 .... Recall thunderstorms—types/characteristics/formation/hazards
2158, 2172, 2173

PLT497 .... Recall transponder—codes/operations/usage
2005

PLT501 .... Recall turbulence—types/characteristics/reporting/corrective actions
2159, 2236

PLT502 .... Recall universal signals—hand/light/visual
2352

PLT509 .... Recall wake turbulence—characteristics/avoidance techniques
2039, 2070, 2071, 2072, 2235

PLT510 .... Recall weather—causes/formation
2171

PLT511 .... Recall weather associated with frontal activity/air masses
2164, 2165, 2312

PLT512 .... Recall weather conditions—temperature/moisture/dewpoint
2169, 2207, 2208

PLT514 .... Recall weather reporting systems—briefings/forecasts/reports
2153, 2154, 2210

PLT517 .... Recall winds associated with high/low-pressure systems
2170

PLT518 .... Recall windshear—characteristics/hazards/power management
2162, 2163

Notes

Notes

Notes

Notes

Notes

Notes

More Sport Pilot Products from ASA

ASA has many other books and supplies for pilots, which are available from an aviation retailer in your area. Need help locating a retailer? Call ASA at **1-800-272-2359** or visit our website (**www.asa2fly.com**). We can also send you the latest ASA Catalog which includes our ***complete*** line of publications and pilot supplies… for all types of pilots and aviation technicians.

Books

	Product Code
Sport Pilot Checkride	ASA-CKRIDE
Say Again, Please: Guide to Radio Communications	ASA-SAP
Sport Pilot Practical Test Standards (Airplane, Weight-Shift Control, Powered Parachute, and Flight Instructor)	ASA-8081-SPORT
Aeronautical Chart User's Guide	ASA-CUG
Dictionary of Aeronautical Terms	ASA-DAT
Airplane Flying Handbook	ASA-8083-3
Pilot's Handbook of Aeronautical Knowledge	ASA-8083-25
Airplane Pilot Combo	ASA-8083-3A-25
Aviation Weather (AC 00-6A)	ASA-AC00-6A
Aviation Weather Services (AC 00-45)	ASA-AC00-45
Aviaton Weather Combo Pak	ASA-AC00-6A-45
FAR/AIM: Federal Aviation Regulations and AIM	ASA-FR-AM-BK
Certified Flight Instructor Test Prep (Includes sport pilot instructor questions)	ASA-TP-CFI

Software and DVDs

ROTAX 912 Engine Introduction: Basic Operation and Maintenance for Pilots and Mechanics	ASA-F2F-ROTAX
Weather to Fly for Sport Pilots, with Paul Hamilton	ASA-F2F-W2F
Sport Pilot: How to Choose the Light-Sport Aircraft That's Right for You	ASA-F2F-SPORT
Preflight a Fixed Wing Light-Sport Aircraft	ASA-F2F-FW-PF
Learn to Fly a Fixed Wing Light-Sport Aircraft	ASA-F2F-FW-L2F
Sport Pilot Checkride DVD	ASA-F2F-CK-DVD
Pro-Flight Library (CD-ROM)	ASA-CD-FL-PRO
Communications Trainer: Say Again, Please	ASA-ESAP
Sport Pilot Prepware	ASA-TW-SPORT
CFI Prepware (includes Sport Pilot Instructor tests)	ASA-TW-CFI
Prepware.com (Internet Knowledge Test Simulator)	ASA-PW.COM

Pilot Supplies

Sport Pilot Logbook	ASA-F2F-LOG
VFR Kneeboard	ASA-KB-1
AirClassics Flightlight	ASA-FL-2
ASA Flight Planner Pad	ASA-FP-2
E6-B Flight Computer	ASA-E6B
CX-2 Pathfinder Flight Computer	ASA-CX-2
CX-2 for Palm PDAs	ASA-CX-2-PALM
Ultimate Rotating Plotter	ASA-CP-RLX
ASA QuickCheck Cards	ASA-QC-1